MIMESIS
INTERNATIONAL

PHILOSOPHY
n. 75

Raffaele Carbone

HUMANISM AND NATURE IN HORKHEIMER AND ADORNO

MIMESIS
INTERNATIONAL

Book published with funds of research project 'Paradoxical Humanism and Environmentalism', PRIN PNRR 2022, Università degli Studi di Napoli Federico II, Dipartimento di Studi Umanistici.

Isbn: 9788869775550
Book series: *Philosophy*, n. 75
Series Issn: 3103-3865

CONTENTS

In memory of Giuseppe Cacciatore

Lui, qui avait son imagination plus pleine et plus étendue, embrassait l'univers comme sa ville: jetait ses connaissances, sa société et ses affections à tout le genre humain: Non pas comme nous, qui ne regardons que sous nous

Montaigne, *Les Essais*, I, 26, ed. by A. Tournon and F. Garavini, Milan: Bompiani, 2012, p. 282.

He, whose imagination was fuller and more extensive, embraced the universe as his city, and distributed his knowledge, his company, and his affections to all mankind, unlike us who look only at what is underfoot

The Complete Essays of Montaigne, trans. by D. M. Frame, Stanford: Stanford University Press, 1958, p. 116.

ACKNOWLEDGEMENTS

I would like to thank all those who contributed to the progress of my work with their suggestions and material assistance. The exchanges of ideas I had with John Abromeit, Maurizio Cambi, Estelle Ferrarese, Katia Genel, Sonja Lavaert, Edoardo Massimilla, Anna Pia Ruoppo, Stefano Petrucciani, Luca Scafoglio and Darrow Schecter were crucial in maturing my reflection on humanism, nature and critical theory.

Finally, I am grateful to Laura Lasagna and James Stunell for proofreading the text and providing me with useful suggestions for improving its style.

INTRODUCTION

In recent decades, objections and criticisms of the idea of humanism have punctuated philosophical debate.[1] Criticisms have come in particular from the viewpoint of environmentalism, but sometimes they promote a possible dialogue between humanism and environmentalist perspectives, which some scholars have even hoped for and envisaged.[2] Nevertheless, it is also essential to consider those internal tensions within the idea of humanism, which have been continuously re-proposed and re-actualised in the Western cultural tradition. Indeed, this study has investigated tensions and paradoxes within humanistic projects, which emerge when their rhetorical and conceptual tools are used to realise a decentralisation of perspective, dethroning the human subject from its self-proclaimed and presumed centrality in the cosmos, to consider human beings from the point of view of the animal or even to rethink their condition in a plural space of living beings and non-living entities.

In this regard it is interesting to refer to the current debate on humanism and specifically on the relationship between 'humanism and environmentalism'. As part of this debate, arguments that

---

1 See Slavica Jakelić, 'Humanism and Its Critics', in *The Oxford Handbook of Humanism*, ed. by Anthony B. Pinn (Oxford: Oxford University Press, 2021), pp. 265–93.

2 See Catherine and Raphaël Larrère, *Du bon usage de la nature: Pour une philosophie de l'environnement* (Paris: Flammarion, 2022 [Paris: Aubier, 1997]), chapter 2: 'Nature et Humanisme', pp. 57–103; Bruno Latour, *Face à Gaïa: Huit conférences sur le nouveau régime climatique* (Paris: Éditions La Découverte, 2015); Dan McKanan, 'Humanism and Environmentalism', in *The Oxford Handbook of Humanism*, pp. 509–30; Rotimi Omosulu and Terzungwe Inja, 'Humanism and Environmentalism: A Dialogue', *Nnamdi Azikiwe Journal of Philosophy*, 11.2 (2019), 87–100.

question 'classical' humanism have been examined, given that its expression in early modernity allegedly entailed a double caesura: between human beings and the other living species, and between Western cultures and other cultures. In this respect, it is worth mentioning the criticisms elaborated in the last century by Sartre,[3] Lévi-Strauss, and Althusser,[4] and more recently by Charles Taylor,[5] Sophie Guérard de Latour,[6] and Camille de Toledo.[7] It should also be noted that in certain contexts, for example in feminist and race theory, the word 'humanism' is used with extreme caution because humanistic ideology is still understood to be responsible for the

3 See Sartre's severe critique of *L'Homme révolté* by Camus. It also takes the form of a critique of a certain humanism that, in the name of high 'human values' and the 'idea of man,' disregards history: this is Camus's 'vieil humanisme', which seeks happiness and assent. See Jean-Paul Sartre, 'Réponse à Albert Camus', in *Les Temps modernes*, 82 (August 1952), in Sartre, *Situations, IV* (Paris: Gallimard, 1964), pp. 90–125 (pp. 114, 116, 117, respectively). See also Sartre's Preface to *Les damné de la terre* (1961): Jean-Paul Sartre, 'Préface', in Frantz Fanon, *Les damnés de la terre* (Paris: La Découverte, 2009), pp. 17–36.

4 Louis Althusser, *Réponse à John Lewis* (Paris: Maspéro, 1973), quoted by Jean-Michel Besnier, 'Un humanisme paradoxal', *Le Télémaque*, 21 (2002), 97–106 (p. 104).

5 Consider, for example, the notion of 'exclusive humanism', which in Taylor designates the humanism of modern times: an 'exclusive', 'secular' humanism that is opposed to all transcendence; a way of being in the world that locates the deepest sources of meaning only in reference to human life, rather than in reference to some reality outside or beyond human life. See Charles Taylor, *A Secular Age* (Cambridge, MA: The Belknap Press of Harvard University Press, 2007), pp. 19–21 *and sequitur*.

6 Sophie Guérard de Latour, 'L'humanisme, une valeur à partager entre différentes cultures. À quelles conditions', *L'Observatoire*, 49.1 (2017), 25–8.

7 Camille de Toledo questions the glorification of the values of humanism ('valeurs de l'humanisme'), which conceals what the Western subject has actually achieved in the course of modernity: the caesura between the one who sees, measures, and orders and the set of objects whose contours he claims to define. According to de Toledo, this separation must be rethought so that the 'lieu du sujet' opens up to the 'réfiés' and the 'rélegués' in a struggle for the rights of the entities of nature. Camille de Toledo, *Une histoire du vertige* (Lagrasse: Verdier, 2023), pp. 108–9.

dehumanisation of women and the practices of racialisation that uphold white supremacy.[8] Nevertheless, a brief review of the history of the idea of humanism shows that a renewed humanism returns periodically.[9]

Sophie Guérard de Latour recently argued that, on the one hand, the humanist ideal, with its precise historical roots in the Italian Renaissance, emphasises the individual's capacity for self-determination and the ability to choose the shape of an existence that embodies universal values; on the other, humanism runs the risk of being (and ultimately becoming) 'a sectarian philosophy as soon as it imposes this vision of things on people who do not subscribe to it'.[10]

---

8 See Ann V. Murphy, 'Corporeal Vulnerability and the New Humanism', *Hypatia*, 26.3 (2011), 575–90 (p. 589). Nevertheless, Butler's descriptions of a 'new bodily ontology" and Cavarero's 'altruistic ontology of relation' (Ibid., p. 578) make it possible to imagine a new humanism: 'The new corporeal humanism, grounded in the ontological fact of vulnerability, dispossession, and exposure, offers a more expansive and inclusive paradigm, one that is still attentive to the differences that mark bodies, and respectful of the radically different ways that vulnerability and dispossession are lived'. Ibid., p. 589. See, in this regard, Judith Butler, *Precarious Life: The Powers of Mourning and Violence* (London: Verso, 2004), pp. 42–3: 'By insisting on a "common" corporeal vulnerability, I may seem to be positing a new basis for humanism. That might be true, but I am prone to consider this differently'.

9 See, for example, Fanon, *Les damnés de la terre*, pp. 225–35; Edward W. Said, *Humanism and Democratic Criticism* (New York: Columbia University Press, 2004). See, in this regard, the book edited by Michael Bryson, *The Routledge Companion to Humanism and Literature* (New York: Routledge, 2022), in particular: Michael Bryson, 'Introduction. The Old Argument: Humanism and Anti-Humanism', pp. 1–9; Deepa Jani, 'Frantz Fanon: Postcoloniality and New Humanism', pp. 31–49; Sauleha Kamal, '"A Different Kind of Humanism": Edward Said's Césairian Critical Humanism', pp. 60–72. On this topics, see G. Cacciatore, 'Nuovo umanesimo e filosofia interculturale', *Humanitas*, 69.4–5 (2014), 584–95.

10 Guérard de Latour, 'L'humanisme, une valeur à partager entre différentes cultures. À quelles conditions?', p. 26. The author develops a severe critique of the humanist ideal, showing that, while it is ethnically situated,

This kind of attitude towards humanism – considered as 'une vieille lune bourgeoise'[11] – is very common nowadays. Nevertheless, the questioning of Western humanism dates back as far as (at least) the twentieth century. Consider, for example, Claude Lévi-Strauss's harsh criticism of a humanism derived from the Renaissance, which emerges in his article 'Jean-Jacques Rousseau, fondateur des sciences de l'homme' (1962):[12]

> We started by cutting man off from nature and establishing him in an absolute reign. We believed ourselves to have thus erased his most unassailable characteristic: that he is first a living being. Remaining blind to this common property, we gave free rein to all excesses. Never better than after the last four centuries of his history could a Western man understand that, while assuming the right to impose a radical separation of humanity and animality, while granting to one all that he denied the other, he initiated a vicious circle. The one boundary, constantly pushed back, would be used to separate men from other men and to claim – to the profit of ever smaller minorities – the privilege of a humanism, corrupted at birth by taking self-interest as its principle and its notion.[13]

In a 1971 text entitled 'Race and Culture', which can be read today in *Le regard éloigné*, Lévi-Strauss returns to the 'separations' and 'frontiers' created by Western humanism. He reaffirms that the isolation of human beings from the rest of nature prepares the ground

---

it has the ambition to set itself up as a universal standard against which any human group can be 'compared, evaluated and ranked'. Ibid., p. 27.

11 François Hartog, *Départager l'humanité. Humains, humanismes, inhumains* (Paris: Gallimard, 2024), p. 316.

12 It is a speech given in Geneva on 28 June 1962 during the ceremonies marking the 250th anniversary of the philosopher's birth, which was later included in *Anthropologie structurale deux* (1973).

13 Claude Lévi-Strauss, 'Jean-Jacques Rousseau, fondateur des sciences de l'homme', in Lévi-Strauss, *Anthropologie structurale deux* (Paris: Plon, 1996 [1962, 1973]), pp. 45–56 (p. 53): 'Jean-Jacques Rousseau, Founder of the Sciences of Man', in Lévi-Strauss, *Structural Anthropology*, Volume 2, trans. by Monique Layton (Chicago: University of Chicago Press, 1983), pp. 33–43 (p. 41).

for the drawing of an ideological line that separates the West from the rest of the world.[14]

This interesting aforementioned analysis indicates that, according to Lévi-Strauss, Western humanism produces two ruptures: it separates man from the rest of nature and draws a border between Western cultures and all other cultures, favouring a hierarchisation of human groups.

Moreover, within this theoretical framework of the second half of the twentieth century, Foucault's delimitation and re-dimensioning of the idea of humanism in the 1960s is well known[15] as is his foreshadowing – in Toussaint's words – of 'an extinction of humanism'[16] starting from the idea that man is not the oldest problem of human knowledge but 'an invention of recent date'.[17]

In present research, there is a considerable number of defenders of humanism, and even within the field of literature and environmental struggles there are voices claiming a humanism in harmony with

---

14 Claude Lévi-Strauss, 'Race et culture', in Lévi-Strauss, *Le regard éloigné* (Paris: Plon, 2001 [1983]), pp. 24–48 (p. 46).

15 'On croit que l'humanisme est une notion très ancienne qui remonte à Montaigne et bien au-delà. Or le mot "humanisme" n'existe pas dans le Littré. En fait, avec cette tentation de l'illusion rétrospective à laquelle on ne succombe que trop souvent, on s'imagine volontiers que l'humanisme a toujours été la grande constante de la culture occidentale. Ainsi, ce qui distinguerait cette culture des autres, des cultures orientales ou islamiques par exemple, ce serait l'humanisme. On s'émeut quand on reconnaît des traces de cet humanisme ailleurs, chez un auteur chinois ou arabe, et on a l'impression alors de communiquer avec l'universalité du genre humain'. Michel Foucault, 'L'homme est-il mort?', entretien avec Claude Bonnefoy (1966), in Foucault, *Dits et écrits I, 1954-1975* (Paris: Gallimard, 2001), pp. 568–72 (p. 568). See Daniel Liotta, 'Foucault: Une introduction à une critique de l'humanisme', *Humanisme*, 293.3 (2011), 35–45.

16 According to Toussaint, the Foucauldian proclamation of an end to humanism through the liberation of a new thought is contradicted in the current age of performance and profit by a condition of subjugation without humanitas. See Stéphane Toussaint, *Humanismes Antihumanismes. De Ficin à Heidegger*, I (Paris: Les Belles lettres, 2008), p. 260.

17 Michel Foucault, *Les mots et les choses* (Paris : Gallimard, 1990), p. 398; *The Orders of Things : An Archeology of the Human Sciences* (New York: Vintage Books, 1994), p. 387.

an environmentalist theory and approach. This kind of humanism calls for action on the environment which should be rooted in the fundamental belief, not of human superiority, but of human responsibility.[18] According to Lewis P. Hinchman, by stressing the critical and historical orientation of the Renaissance humanist tradition, it is possible to 'trace a line of development that runs from Renaissance humanism to both the historical natural sciences and ecology as well as to the humanities, while distinguishing both from sciences in search of abstract, universal laws'.[19] The study of classical languages, cultures and literature is not antiquarian in character, but rather enables us, through the study of the past, to create pathways between worlds that are distant in time, and therefore reflect on their similarities and differences, making it possible to realise a 'fusion of horizons'.[20]

Within the framework of these discussions, some scholars have gone so far as to analyse the existing situation of humanism in the twenty-first century, often on the basis of the debate on humanism that developed in the twentieth century.[21] François Hartog, for example, emphasises the ambivalence with which humanism is viewed in our times ('aujourd'hui, l'humanisme a ou trop bonne ou trop mauvaise presse').[22] In his book, *Humanism Foundations, Diversities, Developments* (2021), on the one hand, Jörn Rüsen states that humanism tends to emphasise the cultural dimension of human beings, which essentially differentiates them from the natural environment; on the other hand, he observes that it cannot be reduced to 'an ideology of human mastery over nature, which

18 https://heritage.humanists.uk/article/a-brief-history-of-humanist-environmentalism/

19 Lewis P. Hinchman, 'Is Environmentalism a Humanism?', *Environmental Values*, 13.1 (2004), 3–29 (p. 9).

20 Hinchman, 'Is Environmentalism a Humanism?', p. 8.

21 See for example *La questione dell'umanismo oggi*, ed. by Annette Hilt, Holger Zaborowski, and Virgilio Cesarone (Macerata: Quodlibet, 2017).

22 François Hartog, 'Lettre sur l'Humanisme et les Humanités', *Anabases. Traditions et réceptions de l'Antiquité*, Études sur la circulation et la réception des savoirs offertes à Pascal Payen. *Actualités et débats*, 35 (2022), 227–43 (p. 228).

ultimately ends in the ecological catastrophe of today'.[23] Francis Wolff, in turn, considers it imperative to plead the cause of a humanism based on the language-reason nexus, as opposed to the limitations and risks of naturocentrism (biocentrism, zoocentrism, and ecocentrism).[24] Furthermore, illustrating the characteristics of ecopsychology, which aims to contribute to the 're-enchantment' of the world, Mohammed Taleb rejects the thesis of certain ecological milieux, according to which the Renaissance was a matrix of the ecological crisis, insofar as it produced anthropocentric humanism. In this regard, according to Taleb, it is appropriate to distinguish between an 'early humanism' (which includes Nicholas of Cusa, Pico della Mirandola, Paracelsus, up to Robert Fludd) that outlines the ideal of homo universalis, a human being who, while retaining his singularity, constitutes himself in a relationship of correspondence and resonance with the cosmos, and a 'second humanism,' the anchor point of capitalist modernity, the embodiment of which is the lonely man, separated from a disenchanted world, who is reduced to his rational personal conscience (his cogito).[25] This second form of humanism, which produces the conception of a nature external to man and society, finds its quintessence in the Baconian idea of the dominion of nature, of the 'vexatio'[26] as the main way to regain

---

23 Jörn Rüsen, *Humanism: Foundations, Diversities, Developments* (London and New York: Routledge, 2021), p. 93: 'Western humanism included the idea that man subjugates and dominates nature. This way of thinking reflects the urban civilization of a city, where human life is no longer close to nature though its original base was grounded on agriculture'. Rüsen then evokes Pico della Mirandola, whose *Oratio* decisively expresses an idea dominant in the modern era: the idea that '[…] God has put the humans into the center of the world and has entitled them to relate the whole creation to themselves, since in humankind the cosmos reflects itself in its sense-bearing order. At the same time the humans are entitled to create the order of their world on their own'. Ibid., p. 94.

24 Francis Wolff, *Plaidoyer pour l'universel. Fonder l'humanisme* (Paris: Arthème Fayard, 2021), pp. 113–21.

25 Mohammed Taleb, 'Psyché, cosmos et âme du monde. Fragments d'histoire de l'écopsychologie', *L'Art du comprendre*, 21, *La nature et son souci. Philosophie et* écologie (2012), 187–201 (p. 196).

26 See Carolyn Merchant, 'Francis Bacon and the "vexations of art": Experimentation as intervention', *The British Journal for the History of*

a harmonious relationship with the natural world, which ultimately means knowing nature through the experimental sciences in order to shape it according to human needs.[27]

Moreover, humanism seems to entail a critical attitude in itself:[28] it is not a naïve and definitive celebration of the humaneness achieved in Western societies and their cultural canon, 'but a reminder that these achievements are permanently undermined by the social, political, and economic conditions of our societies'.[29]

According to various contemporary authors, humanism and humanities are necessary for us. The cultural and literary scholar Mikhail Epstein states:

> To study the human being also means to create humanness itself: every act of the description of the human is, by the same token, an event of one's self-construction. In a wholly practical sense, the humanities create the human, as human beings are transformed by the study of literature, art, history and philosophy: the humanities humanise.[30]

In this regard, Ken Plummer claims the need for a critical humanism grounded in a model of education that cultivates 'an awareness of this extraordinary range of classical and contemporary human stories from all around the world and in all their modes'.[31]

In a recent book, Friedemann Stengel notes that the theme of humanism has a strong moral and normative significance: therefore, it must be approached in a productive manner with an appropriate

---

Science, 46.4 (2013), 551–99.

27 Lusin Bagla, 'La nature entre changement climatique et séquestration du CO_2. De la recherche au pragmatisme', *L'Art du comprendre*, 21 (2012), 203–17 (pp. 209–10).

28 Erich Fromm, 'Humanismus und Psychoanalyse', in Fromm, *Gesamtausgabe IX: Sozialistischer Humanismus und humanistische Ethik* (Munich: DTV, 1999), pp. 3–11 (p. 5).

29 Oliver Kozlarek, *The Critical Humanism of the Frankfurt School as Social Critique*, (Lanham, MD: Lexington Books, 2024), p. 3.

30 Mikhail Epstein, *The Transformative Humanities: A Manifesto* (New York: Bloomsbury, 2012), p. 7.

31 Ken Plummer, *Critical Humanism: A Manifesto for the 21st Century* (Cambridge: Polity Press, 2021), p. 108.

historicisation and account for the current discussion on its scope.[32] The scholar, therefore, invites interpreters and thinkers not to archive or evade this theme, but to address it in the polyvalence of its meanings and contradictions, in order to construct a new approach to the notion which, rather than a historiographical category, constitutes a terrain of discussion and tension.[33]

A brief historical overview of the word 'humanism' and its meanings is outlined below. This brief overview may also be useful in understanding the meaning with which Horkheimer and Adorno use the term 'humanism' in their texts, or in highlighting how, by analysing the reversal of the emancipatory project of the Enlightenment into mere domination (domination of man over nature and domination of man over man), they presuppose the concept of 'humanism' as a founding element of Western civilisation, whose crisis and fall into barbarism they diagnose.

It is the second half of the eighteenth century, in an article by Quesnay that appeared in *Éphémérides du citoyen*,[34] that the word 'humanism' entered the European philosophical-cultural lexicon with the meaning of 'general love of humanity', as a synonym for philanthropy. Nevertheless, it is the nineteenth century that marked a decisive turning point in the concept, primarily through the publication of the philosopher and educationalist Friedrich Immanuel Niethammer's (1766–1848) book *The Dispute of Philanthropinism and Humanism in the Theory of Educational Schooling of Our Time*. Niethammer forged the word 'Humanismus' from the Ciceronian term 'humanitas' and used it as a weapon in a pedagogical battle pitting the classical studies of the *Gymnasium* against the *Realschule*, which focused on the professional training of the bourgeois class. He argued that 'Philanthropinismus'[35] should

32 Friedemann Stengel, *Genealogie des Humanismus. Debatten – Kritik – Neue Perspektiven* (Bielefeld: transcript Verlag, 2024), p. 24.

33 Stengel, *Genealogie des Humanismus*, p. 89.

34 François Quesnay, 'Des sciences et des arts', *Éphémérides du citoyen, ou Bibliothèque raisonnée des sciences morales et politiques*, 1.16 (1765), 241–56 (p. 247).

35 As Sebastiaan van Bommel states, 'with the term "philantropinism", Niethammer referred to a powerful current in Enlightenment pedagogy,

be called 'Animalismus' since it advocates a model of education based not on humanity but on animality: it addresses the animal part of man.[36] Niethammer challenged this 'materialist' model and argued that man is not a simple juxtaposition of animality and rationality, but a unified whole from the two elements,[37] opposing 'Humanismus' to 'Philanthropinismus'. According to the humanistic model, the educational process includes an intellectual and moral formation that, inspired by Kant's thought, has man as its goal with a view to advancing mankind. This type of education is to be realised, through the study of Greek and Latin, in the Gymnasium and at university.[38] Indeed, the core of the conflict between philanthropism and humanism lay in the debate about the teaching of languages, especially classical languages.[39]

While retaining its place in pedagogical debates, the concept of humanism spread to other contexts as early as the 1830s and 1840s. It was enriched with more complex implications to the extent that authors such as Arnold Ruge (1802–1880) reinterpreted it by highlighting a set of social attitudes and positions that it should entail. Along these lines, newspapers also began to attribute new meanings to the concept of humanism in the period before the revolutions of 1848–49.[40]

---

now commonly known under the name *Philantropismus* (or English "Philanthropinism"), which strongly favoured *real*- above classical education'. Sebastiaan van Bommel, *Classical Humanism and the Challenge of Modernity: Debates on Classical Education in 19th-Century Germany* (Berlin: De Gruyter, 2015), p. 112. Humanistic education focuses on 'the life of ideas', on 'liberal knowledge' of 'the true, the good and the beautiful', while 'philantropinist' education focuses chiefly 'on the surrounding material world'. Ibid.

36 Friedrich Immanuel Niethammer, *Der Streit des Philanthropinismus und Humanismus in der Theorie des Erziehungs-Unterrichts unsrer Zeit* (Jena: Frommann, 1808), p. 8.

37 Niethammer, *Der Streit des Philanthropinismus und Humanismus*, p. 67.

38 Hartog, 'Lettre sur l'Humanisme et les Humanités', p. 232.

39 Kristina Bosakova and Martina F. Bykova, 'Hegel and Niethammer on the Educational Practice in Civil Society', *Journal of Philosophy of Education*, 55.1 (2021), 99–125 (pp. 115–16).

40 On these topics see Hans Erich Bödeker, 'Menschheit, Humanismus, Humanität', in *Geschichtliche Grundbegriffe: Historisches Lexikon zur*

Also, towards the middle of the nineteenth century, the term 'Humanismus' entered German usage to denote a historical event, and an intellectual phenomenon associated with the Renaissance.[41]

As has been extensively discussed, the historiographical and philosophical reflection on humanism would develop in the twentieth century, particularly in the first part. In this framework, on the one hand, authors such as Heidegger and Sartre questioned the nexus between the human and its other on the basis of the humanist paradigm; on the other hand, the historiographical work of various scholars and thinkers (Cassirer, Kristeller, Garin and so on) defined the contours and stakes of humanism as a philological and/or a philosophical programme. Some considerations by Ernesto Grassi offer a possible thread to follow in this line of research. Grassi notes that starting from Burckhardt and Voigt and up to Cassirer, Gentile and Garin, the essence of humanism has been identified 'in the rediscovery of man and his immanent values', which triggered Heidegger's polemic against humanism, reduced to a pure and simple consideration of everything from man and his perspective. By contrast, according to Grassi, '[o]ne of the central problems of Humanism, however, is not man, but the question of original context,

---

politisch- sozialen Sprache in Deutschland, ed. by Otto Brunner, Werner Conze, and Reinhart Koselleck (Stuttgart: Klett-Cotta, 1982), vol. 3, pp. 1063–1128; Georg Bollenbeck, *Bildung und Kultur: Glanz und Elend eines deutschen Deutungsmusters* (Frankfurt am Main: Insel, 1994), pp. 142–55; Heidi Hakkarainen, 'Contagious Humanism in Early Nineteenth-Century German-Language Press', *Contributions to the History of Concepts*, 15.1 (2020), 22–46 (pp. 23–4).

41 Karl Hagen, *Deutschlands literarische und religiöse Verhältnisse im Reformationszeitalter. Mit besonderer Rücksicht auf Wilibald Pirkheimer*, 3 vols (Erlangen: Palm, 1841-1844), Vol. 1, p. 79 and *passim*; Georg Voigt, *Die Wiederbelebung des classischen Alterthums oder das erste Jahrhundert des Humanismus*, 2 vols. (Berlin: Reimer 1859); Jacob Burckhardt, *Die Kultur der Renaissance in Italien. Ein Versuch*, chapter III, 4: 'Der Humanismus im 14. Jahrhundert', in Burckhardt, *Das Geschichtswerk* (Frankfurt am Main: Zweitausendeins, s.d.), Vol. I, p. 476–80. See Mario Todte, *Georg Voigt (1827–1891): Pionier der historischen Humanismusforschung* (Leipzig: Leipziger Universitätsverlag, 2004); Rémi Brague, *Le propre de l'homme* (Paris: Flammarion, 2015).

the horizon, or "openness" in which man and his world appear',[42] and the specific way in which humanism addresses these problems consists in the analysis and interpretation of language, understood as the historical horizon in which things appear.[43]

In this debate, it is therefore appropriate to examine some texts by Horkheimer and Adorno, which, as will be shown, explicitly address not only the problem of the relationship between man and nature, but also that of humanism as a fundamental orientation of Western civilisation, highlighting its internal tensions and possible reinterpretations. In this regard, it should be emphasised that these themes are addressed in individual texts by each of the two authors separately, but also in the book they wrote together, and in any case their reflections are the result of a decade of discussion and collaboration.

The aim of this book is to examine how reflections on the meaning and relevance of humanism emerge in the writings of Horkheimer and Adorno, and to what extent these reflections include considerations on the relationship between human beings and the specific natural and environmental contexts that they transform according to their needs and plans. The question is legitimate insofar as in some texts both authors face explicitly the question of possible 'reactualisations' of humanism. Horkheimer, for example, examines the neo-humanism of his time in an article published in 1938, in which he discusses Siegfried Marck's book, *Der Neuhumanismus als politische Philosophie*,[44] while Adorno discusses the meaning of 'humanism' in the text of a conference, entitled 'Über Technik und Humanismus'.[45] Moreover, their questioning of what it means

---

42 Ernesto Grassi, *Heidegger and the Question of Renaissance Humanism: Four Studies* (Binghamton: Center for Medieval and Early Renaissance Studies, 1983), p. 17.

43 Grassi, *Heidegger and the Question of Renaissance Humanism*, pp. 26–7.

44 Siegfried Marck, *Der Neuhumanismus als politische Philosophie* (Zürich: Aufbruch, 1938). See Max Horkheimer, 'Die Philosophie der absoluten Konzentration', in Horkheimer, *Gesammelte Schriften* (henceforth *GS*), ed. by Alfred Schmidt and Gunzelin Schmid Noerr (Frankfurt am Main: Fischer, 1985–1996), vol. 4 (2nd edition: 2009), pp. 295–307

45 Theodor W. Adorno, 'Über Technik und Humanismus', in Adorno, *Gesammelte Schriften in 20 Bänden* (henceforth *GS*), ed. by Rolf

to be human (or rather their reflections on the 'inhuman') includes a detailed examination of the complex relationship that human beings establish with the natural environment.

Finally, this book does not aspire to provide an exhaustive treatment of two such complex and multi-layered concepts and their interrelationship in the works of Horkheimer and Adorno. It aims to address these concepts primarily in their main works in order to highlight their weight and the role they play in the development of their theses on the crisis of Western societies and on the complex dialectic between subject and nature; the intent is also to identify useful theoretical elements for addressing current discussions on the concept of nature in its contrasting relationship with humanism.

Tiedemann, with the collaboration of Gretel Adorno, Susan Buck-Morss and Klaus Schultz (Frankfurt am Main: Suhrkamp, 1970–1997), Vol. 20.1: *Vermischte Schriften I. Theorien und Theoretiker – Gesellschaft, Unterricht, Politik*, 5th edition (Frankfurt am Main: Suhrkamp, 2023), pp. 310–17.

I.
HORKHEIMER AND ADORNO ON HUMANISM AND NATURE IN THE 1930S

1. *Introduction: Horkheimer and Adorno on humanism and nature at the beginnings of the critical theory project*

In Horkheimer and Adorno, the findings indicate that there is a critical attitude towards the traditional concepts of 'humanism' and 'nature'. Frankfurt thinkers tackle these concepts of Western philosophy but with the aim of showing their internal limits and their dialectical connections. Moreover, they also consider the possibility of re-elaborating these concepts in the light of the specific needs of critical theory.

It will be argued that a reflection on humanism in Horkheimer and Adorno implies a thematisation of the dialectical relation which connects men to nature, men to other living beings. One of the central problems of Horkheimer and Adorno's theory – in their individual works and in jointly-written texts – is the problem of domination. This term refers to human beings' response to a hostile nature which they fear. Domination very soon reveals itself as control over man's internal nature (primordial impulses) and over external nature. Internal nature is the 'the animal prehistory [*der tierischen Vorzeit*]'.[1] Domination appears within a dialectical process whereby men, in order to face the dangers of external nature, have learned to

1 Max Horkheimer and Theodor W. Adorno, *Dialektik der Aufklärung. Philosophische Fragmente*, in Horkheimer, *GS* 5: *'Dialektik der Aufklärung' und Schriften 1940-1950*, ed. by Gunzelin Schmid Noerr (Frankfurt am Main: Fischer, 2004 [1987]), p. 217; *Dialectic of Enlightenment: Philosophical Fragments*, trans. by Edmund F. N.

dominate the internal nature. 'Civilization is the triumph of society over nature – a triumph which transforms everything into mere nature [*in bloße Natur*]'.[2] At the same time, their analyses also stress that humans' claim to the right to dominate the natural world turns out to be an illusion, a paradoxical dominion. Indeed, in 'Trying to Understand *Endgame* [Versuch, das Endspiel zu verstehen]' Adorno affirms: 'No statement sounds more absurd than this most rational of statements, which reduces "everything" to an "only", the mirage of a world that can be dominated anthropocentrically [*dem Trugbild der anthropozentrisch beherrschbaren Welt*]'.[3]

Horkheimer and Adorno challenge the idea that nature can be entirely controlled. It is the bourgeois illusion which emerges in a complex relationship of humans to nature, analysed particularly in the *Dialectic of Enlightenment*. In the wake of Marx – particularly Marx's early writings and *The German Ideology*, published in its entirety in 1932[4] – they reinscribe human activities within broader

---

Jephcott (Stanford: Stanford University Press, 2002), 'Elements of Anti-Semitism: Limits of Enlightenment', VI, p. 154.

2 Horkheimer and Adorno, *Dialektik der Aufklärung*, p. 216; *Dialectic of Enlightenment*, 'Elements of Anti-Semitism: Limits of Enlightenment', V, p. 153. 'The Jews appeared to have successfully achieved what Christianity had attempted in vain: the disempowerment of magic [*die Entmächtigung der Magic*] by means of its own strength, which, as worship of God, is turned against itself. They have not so much eradicated the adaptation to nature as elevated it to the pure duties of ritual. [...] They are pronounced guilty of what, as the first citizens, they were the first to subdue in themselves: the susceptibility to the lure of base instincts, the urge toward the beast and the earth, the worship of images'. Horkheimer and Adorno, *Dialektik der Aufklärung*, p. 216; *Dialectic of Enlightenment*, p. 153.

3 Theodor W. Adorno, *GS* 11: *Noten zur Literatur*, ed. by Rolf Tiedermann, with the collaboration of Gretel Adorno, Susan Buck-Morss and Klaus Schultz (Frankfurt am Main: Suhrkamp, 2022 [2003]), 'Versuch, das Endspiel zu verstehen', pp. 281–321 (pp. 318–9); *Notes to literature*, ed. by Rolf Tiedemann, trans. by Shierry Weber Nicholson (New York: Columbia University Press, 2019 [1991]), 'Trying to Understand *Endgame*', pp. 237– 67 (p. 265).

4 'Marx admitted no absolute division between nature and society, and hence no fundamental methodological distinction between the natural sciences and historical science. [...] An 'opposition between nature and

natural parameters. Furthermore, in *The German Ideology*, Marx and Engels write: 'History can be contemplated from two sides; it can be divided into the history of nature and the history of mankind. However, the two sides are not to be divided off; as long as men exist the history of nature and the history of men are mutually conditioned'.[5]

Nevertheless, Horkheimer and Adorno refuse to place society entirely within the natural order, just as they refuse to reduce everything to social production. It is important to bear in mind that Western societies are based on the domination of nature and that, in order to engage with this issue, it is necessary to conceptualise both the social organisation of domination and the natural elements over which it is exercised.[6] Adorno proposes, in particular, to thematise the social structure of domination, through which nature has come to repress itself in human beings. As Jean-Baptiste Vuillerod writes, this implies that Adorno's naturalism does not cancel the distinction between the social and the natural by reducing the first to the second (or vice versa).[7] More precisely, as Deborah Cook observes, according to Adorno, 'human history is linked inextricably to both our own internal, instinctual, nature and non-human nature'. Conversely, 'nature is historical, not just because nature evolves and constantly changes, but because it has been profoundly – often negatively – affected by human history'.[8] In brief, the reflection on natural history emphasises the dynamic, and potentially devastating, interaction between nature and history.

---

history' is created by ideologists in that they exclude from history the productive relation of men to nature'. Alfred Schmidt, *The Concept of Nature in Marx*, trans. by Ben Fowkes (London: Verso, 2014), p. 49.

5 Karl Marx and Friedrich Engels, *Deutsche Ideologie*, in *Marx-Engels Gesamtausgabe*, Vol. 5, Part 1, Berlin, 1932, p. 567. It is a textual variant which was not included in the final version of *The German Ideology*, as printed in the more accessible 1953 Berlin edition.

6 On this point, see Jean-Baptiste Vuillerod, *Theodor W. Adorno. La domination de la nature* (Paris: Éditions Amsterdam, 2011), pp. 18–19.

7 Vuillerod, *Theodor W. Adorno. La domination de la nature*, p. 19.

8 Deborah Cook, *Adorno on Nature* (London and New York: Routledge, 2014 [2011]), p. 1.

It should be noted that Horkheimer and Adorno's critical reflection on the destructive and self-destructive relation of men to nature goes back to 1930, decades before the environmental movement emerged in the 1960s. As is well known, at the beginning of the decade Horkheimer became director of the Institute for Social Research and left his mark on the research that would be carried out there. On 24 January 1931, he delivered his inaugural speech for the chair of social philosophy and directorship of the Institute, entitled 'The Present Situation of Social Philosophy and the Tasks of an Institute for Social Research'. This lecture, and the numerous essays written by Horkheimer in the early 1930s – also as the result of internal dialogue within the institute among its various members – would develop a conception of interdisciplinary social research on materialistic foundations that was meant to guide the activities of the institute during Horkheimer's tenure as director. In this period Horkheimer would publish (under the pseudonym Heinrich Regius) a book entitled *Dämmerung. Notizen in Deutschland* (Zurich, 1934), which will be examined in the third paragraph of this chapter.

For his part, at the beginning of the 1930s, Adorno elaborates his early philosophical texts, among which 'The Idea of Natural History' will be analysed in the next paragraph. The following paragraphs will now examine these two texts with a view to highlighting the respective positions of the two authors on the connection between human beings and the natural environment, or between history and nature, in the first half of the 1930s.

2. *The problem of entwinement between nature and history: Adorno's 'The Idea of Natural History' (1932)*

Theodor W. Adorno originally criticized the malady of human society with the ultimate aim of reshaping that relationship in more mutually beneficial ways in a 1932 essay, 'The Idea of Natural History'. He presented this talk on 15 July 1932, as a lecture at a meeting of the Frankfurt section of the Kant Society. Together with the other two philosophical texts from this period – another lecture, 'The Relevance of Philosophy', delivered on 8 May 1931, and the

'Theses on the Language of the Philosopher' – 'The Idea of Natural History' was written between the drafting – which began in the late 1920s – of a work on Kierkegaard that was to form his habilitation thesis and the publication, in 1933, of a revised version of the same manuscript, entitled *Kierkegaard: Construction of the Aesthetic*. In 'The Idea of Natural History' Adorno elaborates a critique of Heidegger's ontology, particularly of his concept of 'historicity', by situating him in the context of contemporary German philosophy's attempt to solve the problem of historicist relativism. Starting from such a criticism, Adorno examines this problem as that of conceiving the unity of history and nature.[9]

Specifically, in this article, criticising the dualism of nature and history posited by Neo-Kantianism and phenomenology and the idealism of Heidegger's ontology insofar as it affirms the priority of reason in gaining insight into the totality,[10] Adorno advanced the project of showing that human history is always also natural history and that non-human nature is entwined with history. He writes that philosophy has to overcome the division of the world into two great realms, nature and spirit or nature and history, and that this division has to be replaced by 'a formulation that achieves in itself the concrete unity of nature and history'.[11] According to Adorno, the concepts of nature and history are not exclusive, but mutually determining: 'each provided the key for the demythification of the other'.[12] Spengler, as Adorno would say a few years later, is mistaken in pushing aside nature with which humans struggle in history. It is no coincidence that, in his thinking, history is 'transformed into a

9 See Robert Hullot-Kentor, 'Introduction to T. W. Adorno's "The Idea of Natural-History"', in Id., *Things Beyond Resemblance: Collected Essays on Theodor W. Adorno* (New York: Columbia University Press, 2006), pp. 234–51.

10 Theodor W. Adorno, 'Die Idee der Naturgeschichte', in *GS* 1: *Philosophische Frühschriften* (Frankfurt am Main: Suhrkamp, 1990), pp. 345–66 (pp. 346–54); 'The Idea of Natural History', *Telos: Critical Theory of the Contemporary*, 60 (1984), 111–24, (pp. 112–16).

11 Theodor W. Adorno, 'Die Idee der Naturgeschichte', p. 354; 'The Idea of Natural History', pp. 116–17.

12 Susan Buck-Morss, *The Origin of Negative Dialectics* (New York: The Free Press, 1977), p. 54.

second nature, as blind, closed, and fateful as any vegetable life'.[13] In this vein, Adorno puts forward 'the idea of natural history' as an 'ontological reorientation of the philosophy of history'.[14] According to him, in order to overcome the one-sidedness of subjective idealism and limits of the existential ontology, it is necessary to think of the 'historical being' as a 'natural being' and 'to comprehend nature as a historical being'.[15] As Deborah Cook summarizes Adorno's argument,

> the separation of history from nature is deceptive because, throughout our history, we have engaged with the natural world in productive and reproductive activities for the purpose of self-preservation. Conversely, nature is historical because it constantly develops and changes, both ontogenetically and phylogenetically. Moreover, nature is always also bound up with the historically and socially conditioned concepts and practices that we use to grasp and manipulate it.[16]

Here, Adorno follows Marx and Engels' thesis developed in *The German Ideology*: it is not possible to separate nature from history or history from nature.[17] In *Negative Dialectic*, by interpreting the

13 Theodor W. Adorno, 'Spengler nach dem Untergang', in Adorno, *Prismen*, *GS* 10.1: *Kulturkritik und Gesellschaft I, Prismen – Ohne Leitbild*, 7th edition (Frankfurt am Main, Suhrkamp, 2018), pp. 47–71 (p. 67); 'Spengler after the Decline' (1938), in Adorno, *Prisms*, trans. by Shierry Weber Nicholsen and Samuel Weber (Cambridge, MA: The MIT Press, 1983), 51–72 (p. 69).

14 Adorno, 'Die Idee der Naturgeschichte', p. 355; 'The Idea of Natural History', p. 117.

15 Adorno, 'Die Idee der Naturgeschichte', pp. 354–5; 'The Idea of Natural History', p. 117.

16 Cook, *Adorno on Nature*, p. 17.

17 'We know only a single science, the science of history. One can look at history from two sides and divide it into the history of nature and the history of men. The two sides are, however, inseparable; the history of nature and the history of men are dependent on each other so long as men exist. The history of nature, called natural science, does not concern us here; but we will have to examine the history of men, since almost the whole ideology amounts either to a distorted conception of this history or to a complete abstraction from it. Ideology is itself only one of the aspects of this history'. Karl Marx and Friedrich Engels, *Die deutsche Ideologie*,

aforementioned quotation of *The German Ideology*, Adorno affirms that the traditional antithesis between nature and history is true in one respect and false in another. The opposition is 'true insofar as it expresses what happened to the natural element', namely that so-called first nature has been occluded to such a degree that what now appears to be natural is actually social in character. On the other hand, the antithesis is false to the extent that 'by means of conceptual reconstruction, it apologetically repeats the concealment of history's natural growth by history itself'.[18]

According to Marx and Engels, a thing called nature that lies outside the domain of human history no longer exists anywhere. They stress this point criticising Feuerbach's conception of the 'sensuous world', reduced to the level of mere contemplation or pure feeling. Concerning Feuerbach they write:

> He does not see that the sensuous world around him is not a thing given direct from all eternity, remaining ever the same, but the product of industry and of the state of society; and, indeed [a product] in the sense that it is a historical product, the result of the activity of a whole succession of generations, each standing on the shoulders of the preceding one, developing its industry and its intercourse, and modifying its social system according to the changed needs. Even the objects of the simplest 'sensuous certainty' are only given him through social development, industry and commercial intercourse. The cherry-tree, like almost all fruit-trees, was, as is well known, only a few centuries ago transplanted by *commerce* into our zone, and therefore only *by* this action of a definite society in a definite age has it become 'sensuous certainty' for Feuerbach.[19]

---

in Marx and Engels, *Werke*, vol. 3 (Berlin: Dietz Verlag, 1978), p. 18; *The German Ideology*, in *Collected Works*, vol. 5, trans. by W. Lough, C. Dutt & C. P. Magill (London: Lawrence & Wishart, 1974, Electric Book, 2010), pp. 28–9.

18 Theodor W. Adorno, *Negative Dialektik*, in Adorno, *Gesammelte Schriften*, edited by Rolf Tiedemann, with the collaboration of Gretel Adorno, Susan Buck-Morss und Klaus Schultz, vol. 6: *Negative Dialektik. Jargon der Eigentlichkeit*, 8th edition (Frankfurt am Main: Suhrkamp, 2018), p. 351; *Negative Dialectics*, trans. by E. B. Ashton (New York: Taylor & Francis e-Library, 2004), p. 358.

19 Marx and Engels, *The German Ideology*, p. 39.

Feuerbach emphasised man's opposition to the rest of nature but remained on an abstract level. For him, nature as a whole is an unhistorical, homogeneous substrate, while Marx aims to dissolve this homogeneity in a dialectic of subject and object. According to Marx, nature is both a fundamental element of human practice and the totality of everything that exists.[20]

In his well-known essay-manifesto of 1937, Horkheimer himself would affirm the same idea with different words:

> The world which is given to the individual and which he must accept and take into account is, in its present and continuing form, a product of the activity of society as a whole. The objects we perceive in our surroundings – cities, villages, fields, and woods – bear the mark of having been worked on by man. It is not only in clothing and appearance, in outward form and emotional make-up that men are the product of history. Even the way they see and hear is inseparable from the social life-process as it has evolved over the millennia. The facts which our senses present to us are socially preformed in two ways: through the historical character of the object perceived and through the historical character of the perceiving organ. Both are not simply natural; they are shaped by human activity, and yet the individual perceives himself as receptive and passive in the act of perception.[21]

The discussion now returns to Adorno's 1932 article. The idea of natural history is characterized as methodological proposal for a new form of critical historiography that aims to overcome the conventional idealistic conceptualisation of 'timeless nature' and 'historical progress'. It intends, on the one hand, to explore the way in which historical events, considered precisely as time-bound and singular, reveal themselves to be operations of nature too and, on the other, to criticise the idea of nature as an emblem of timelessness to the extent that it is intertwined with social practices and rational

20 See Schmidt, *The Concept of Nature in Marx*, p. 27.

21 Max Horkheimer, 'Traditionelle und kritische Theorie', in Horkheimer, *GS* 4: *Schriften 1936-1941*, ed. by Alfred Schmidt, 2nd edition (Frankfurt am Main: Fischer, 2009), pp. 162–216; 'Traditional and Critical Theory', in Max Horkheimer, *Critical Theory: Selected Essays*, trans. by Matthew J. O'Connell and others (New York: Continuum, 2002), pp. 188–243 (p. 200).

inquiries of humans.[22] According to Max Pensky, in the light of Adorno's later work, the concept of *Naturgeschichte* also implies 'a pervasive and ultimately paralysing sense of dread and helplessness in the face of a homogenous and virtually irresistible history of domination, and a corollary sense of capitulation at the vision of world history as continuous catastrophe'.[23] In summary, such a concept, warning against rigid dichotomies, notably the illusion of being able to conceive of nature in itself, separate from human beings, and the methodological error of conceiving human facts as a separate realm, also stimulates a humanistic reflection on domination as an almost inescapable configuration of complex human-nature relations. It is interesting to note that the expression 'natural history' also appears in an article published by Horkheimer in 1933, 'Materialism and Metaphysics', in a context in which the author points out that, despite science's attempts to precisely determine the subjective element present in its investigations, subject and object can never be sharply separated, nor can knowledge be equivalent to its object without residue. In this regard, Horkheimer argues that nature is a factor that resists (*widerstrebenden*) the identity pursued in the sciences. He then states the following: 'Physics is an abstract product of human action and may be related to future experience only as a highly conditional hypothesis, not as the full reflection of a supposed essence of natural history [*Naturgeschichte*]'.[24]

In recoding the concepts of 'nature' and 'history' outside the horizon of their usual meanings, Adorno refers to the pre-Marxist works of Lukács (*The Theory of the Novel*) and Benjamin' *The Origin of German Tragic Dram*.[25] Adorno particularly emphasises

---

22 See Max Pensky, 'Natural History: the Life and Afterlife of a Concept in Adorno', *Critical Horizons*, 5 (2004), 227–58 (p. 231).

23 Pensky, 'Natural History', p. 229.

24 Max Horkheimer, 'Materialismus und Metaphysik', in *Gesammelte Schriften*, vol. 3: *Schriften: 1931-1936*, ed. by Alfred Schmidt, 2nd edition (Frankfurt am Main: Fischer, 2009), pp. 70–105 (p. 88); 'Materialism and Metaphysics', in Horkheimer, *Critical Theory: Selected Essays*, pp. 10–46 (pp. 28–9).

25 Adorno, 'Die Idee der Naturgeschichte', pp. 355; 'The Idea of Natural History', p. 117.

Lukács' concept of 'second nature', which Lukács uses to describe 'the world of convention', the world of things created by humans (and yet lost to them),[26] more precisely; the sphere of culture and social practices which, originally social constructs, become so habitual that they appear natural. Lukács uses the concept of second nature in *The Theory of the Novel* in order to illustrate how the novel as a literary form mirrors the cultural superstructure of the modern society that produced it. Indeed, the nineteenth century novel is entirely governed and permeated by convention, 'the full realisation of the concept of a "second nature"; a quintessence of meaningless laws in which no relation to the soul can be found'.[27] It means that the culture of the society that has produced this novel has also become merely conventional.

Consider now how Lukács describes 'the world of convention':

> It is a second nature [*eine zweite Natur*], and, like nature (first nature), it is determinable only as the embodiment of recognised but senseless necessities and therefore it is incomprehensible, unknowable in its real substance.[28]

In this paragraph Lukács refers to 'the process of *becoming-man*' [*Menschwerdung*] of the soul, which occurs in the social sphere ('its activity among men', *ihrer Tätigkeit unter den Menschen*).[29] This process implies that – as Hegel points out – instinctual acts become human habits in the specific sense that they must be willed

26 Adorno, 'Die Idee der Naturgeschichte', pp. 355; 'The Idea of Natural History', p. 117.

27 Georg Lukács, *Die Theorie des Romans: Ein geschichtsphilosophischer Versuch über die Formen der großen Epik*, 9th Edition (Darmstadt and Neuwied: Luchterhand, 1971), p. 99; *The Theory of the Novel: A Historico-Philosophical Essay on the Forms of Great Epic Literature*, trans. by Anna Bostock (Cambridge, MA: The MIT Press 1971), p. 113.

28 Georg Lukács, *Die Theorie des Romans*, p. 53; *The Theory of the Novel*, p. 62.

29 Georg Lukács, *Die Theorie des Romans*, p. 53; *The Theory of the Novel*, p. 62.

and learned:[30] it is a wide-ranging training in which the human body ceases to be mere animal organism and is transformed into spirit.[31] Nevertheless, Lukács does not limit himself to describing the process through which the various norms, by regulating animal functions, constitute themselves as cultural practices. He highlights the omnipotence of the world of the habits, of conventions (its 'all-embracing power', *Allgewalt*), its pervasiveness ('a world which is present everywhere in a multiplicity of forms too complex for understanding'), its laws manifesting in their necessary evidence to the cognisant subject and yet devoid of meaningfulness to its eyes.[32] Indeed, the social and intersubjective 'process of *becoming-man*' – which is a humanistic issue, even evoking an educational program – is reversed into its opposite: human products and relations in their historical development no longer appear to be thus, but rather, losing their human meaning in their dependence of 'laws' that regulate 'human-constructed structures [*Menschengebilde*]',[33] seem artificially natural. In this way, here Lukács highlights 'man's alienation from his own constructs [*Entfremdung zwischen dem Menschen und seinen Gebilden*]' and characterises the first nature – which designates a set of laws for pure cognition or the source of comfort to pure feeling – as 'the historico-philosophical objectivation' of such an alienation,[34] produced by a meaningless mass of protocols, practices, institutions and objects in which men do not recognise themselves. The second nature, the world of convention, made of senseless necessities and therefore unknowable in its actual substance – as Adorno states –, is a world of 'estranged things [*fremd gewordenen Dinge*] that cannot be decoded but encounters us as

30 See Hegel's *Philosophy of Mind*, trans by W. Wallace and A. V. Miller, revisions and commentary by M. J. Inwood (Oxford: Clarendon Press, 2007), § 396, pp. 53–7.

31 See Cat Moir, 'Second Nature and the Critique of Ideology in Hegel and the Frankfurt School', in *Hegel and the Frankfurt School*, ed. by Paul Giladi (New York – Abingdon, Oxon: Routledge, 2021), pp. 115–39 (p. 117).

32 Lukács, *Die Theorie des Romans*, p. 53; *The Theory of the Novel*, p. 62.

33 Lukács, *Die Theorie des Romans*, pp. 54–5; *The Theory of the Novel*, pp. 63–4 (translation amended).

34 Lukács, *Die Theorie des Romans*, p. 55; *The Theory of the Novel*, p. 64.

ciphers'.[35] This world does not appear as the product of a collective creation, of the creative activity of human subjects (that creative nature that emerges in lyric poetry, capable of connecting humans with the archaic act of creation and, consequently, of revealing the factual transience of second nature);[36] its laws and inner mechanism seem to be natural because they are opaque and not accessible to human understanding. Such a perception also distorts modern human sensitivity to nature:

> Estrangement from nature [*Die Fremdheit der Natur*] (the first nature), the modern sentimental attitude to nature, is only a projection of man's experience of his self-made environment as a prison instead of as a parental home.[37]

According to Adorno, Lukács perceives the metamorphosis of the historical (i.e. what has been, our understanding of the past) into nature through the concept of 'petrified' history, he conceives of second nature as history that is blocked, paralysed, 'a charnel-house of rotted interiorities',[38] but he considers the possibility of a spiritual reawakening only as a metaphysical one.[39] He can only conceive of this revival from the perspective of totality, of a re-membered whole, projected into an eschatological future, in terms of revelation.[40] In this regard, Adorno thinks it is necessary to recall some of Benjamin's concepts, as they are complementary to Lukács's and therefore allow him to complete his vision of the dialectical connection between history and nature.

Adorno quotes, in particular, a passage from *The Origin of German Tragic Drama*: 'In nature they [the allegorical poets] saw

---

35 Adorno, 'Die Idee der Naturgeschichte', p. 356; 'The Idea of Natural History', p. 118.

36 On this point, see Moir, 'Second Nature and the Critique of Ideology in Hegel and the Frankfurt School', p. 127.

37 Lukács, *Die Theorie des Romans*, p. 55; *The Theory of the Novel*, p. 64.

38 Lukács, *Die Theorie des Romans*, quoted by Adorno: 'Die Idee der Naturgeschichte', p. 357; 'The Idea of Natural History', p. 118.

39 Adorno, 'Die Idee der Naturgeschichte', p. 357; 'The Idea of Natural History', p. 118.

40 See Gilles Moutot, *Essai sur Adorno* (Paris: Payot, 2010), pp. 56–7.

eternal transience, and here alone did the saturnine vision of this generation recognize history'.[41]

> The deepest point – Adorno comments – where history and nature converge lies precisely in this element of transience. If Lukács demonstrates the retransformation of the historical, as that which has been, into nature, then here is the other side of the phenomenon: nature itself is seen as transitory nature, as history.[42]

Benjamin allows us to understand that nature, as creation, bears the mark of 'transience [*Vergänglichkeit*]': nature itself is 'transitory' [*vergänglich*]' and therefore includes a historical character. What is considered to be historical refers to the natural element inherent therein.[43]

> The basic quality of the transience of the earthly [*Die Grundbestimmung der Vergänglichkeit des Irdischen*] signifies nothing but just such a relationship between nature and history: all being or everything existing is to be grasped as the interweaving of historical and natural being [*als Verschränkung von geschichtlichem und naturhaftem Sein*].[44]

In summary, Benjamin affirms the convergence of nature and history in the moment of transitoriness: 'The word "history" stands written on the countenance of nature in the characters of transience. The allegorical physiognomy of nature-history, which is put on stage in the *Trauerspiel*, is present in reality in the form of ruin'.[45] It means, according to the 'radical natural-historical thought', that 'everything

41 Walter Benjamin, *The Origin of German Tragic Drama*, trans. by John Osborne (London: Verso, 1977), p. 179.

42 Adorno, 'Die Idee der Naturgeschichte', p. 357; 'The Idea of Natural History', p. 119.

43 Adorno, 'Die Idee der Naturgeschichte', p. 359; 'The Idea of Natural History', p. 120.

44 Adorno, 'Die Idee der Naturgeschichte', p. 360; 'The Idea of Natural History', p. 121.

45 Benjamin, *The Origin of German Tragic Drama*, p. 177.

existing transforms itself into ruins and fragments'.[46] Therefore, if Lukács' *Theory of Novel* casts light on the fragmented relationships (between one individual and another, as well as between individuals and nature) of modern times in order to uncover their characteristic pathologies and concealing techniques,[47] Benjamin's *The Origin of German Tragic Drama* leads to the idea of natural history which discloses the shattered and transitory character of both natural things and human events. According to Deborah Cook, such an entwinement between nature and history implies damage inflicted on natural processes by human constructs, particularly 'their subordination to ends extrinsic to them'.[48] Moreover, for Adorno, the problem of natural history enables the perception of nature itself as dynamic and the restoration of humanity to its natural constitution, neutralising the idealistic principle of purity of thought.[49] This is possible insofar as he applies 'nature' and 'history' 'as mutually determining, mutually corrective concepts'.[50]

In his later lectures on *History and Freedom* (1964–65), Adorno would again evoke his 1932 conference highlighting the critical character of the concept of natural history. In these lectures, he refuses the ontologists' claim that we have now overcome the divergence of nature and history. In this regard, Adorno outlines the ideological character of ontology to the extent that it affirms a reconciliation in the mind, while no reconciliation is achieved in reality.[51] The question of the essence of nature is obscured by the totality of what is postulated. Furthermore, nature considered as the

46 Adorno, 'Die Idee der Naturgeschichte', p. 360; 'The Idea of Natural History', p. 121.

47 See Pensky, 'Natural History: the Life and Afterlife of a Concept in Adorno', p. 233.

48 See Cook, *Adorno on Nature*, p. 18.

49 See Luca Scafoglio, *Il disagio dell'astrazione. Una lettura a partire da Adorno* (Naples: La Città del Sole, 2021), pp. 117–19.

50 Buck-Morss, *The Origin of Negative Dialectics*, p. 57.

51 Theodor W. Adorno, *Zur Lehre von der Geschichte und von der Freiheit* (1964/1965), ed. by Rolf Tiedemann (Frankfurt am Main: Suhrkamp, 2006), pp. 177–8; *History and Freedom: Lectures 1964–1965*, edited by Rolf Tiedemann, trans. by Rodney Livingstone (Cambridge and Malden, MA: Polity Press, 2006), p. 123.

first absolute and immediate thing, as opposed to its mediations, ends up being a function of the act of postulating, thanks to which this unpostulated thing is unmasked as its opposite, as something that has been produced. Therefore, one should try to consider nature in its entirety, and everything that is considered nature, as history. 'Conversely, however, everything historical has to be regarded as nature because, thanks to its own violent origins [*Gesetztheit*], it remains under the spell of blind nature, from which it struggles to dissociate itself'.[52]

In this way, on the one hand, history becomes a means of demystification that unmasks the appearance of 'naturalness' enveloping certain social phenomena: by revealing their historicity, it brings out the possibility of transforming them. On the other hand, the reference to nature as fragmentation and ruin – together with the awareness of history as a sphere of discontinuity (in the sense that it contains both disparate facts and circumstances and structural heterogeneity)[53] – allows us to demystify the impressive narratives of history as a march towards progress. History continues to advance on the ruins of nature; in a certain sense, it is nothing more than the destruction of the individual, concrete, mortal beings who are its fabric.[54]

The idea of nature's historicity – linked to attempts to incorporate it into culture – emerges, then, in a different context, in the attention paid to the discovery of dinosaur remains, on which Adorno reflects in *Minima Moralia* (1951). Here Adorno interprets the 'desire for the presence of the most ancient' as 'a hope that animal creation might survive the wrong that man has done it, if not man himself, and give rise to a better species, one that finally makes a success

52 Adorno, *Zur Lehre von der Geschichte und von der Freiheit*, p. 179; *History and Freedom*, p. 124.

53 Adorno, 'Die Idee der Naturgeschichte', pp. 361–2; 'The Idea of Natural History', p. 122.

54 On this point, see Jacques-Olivier Bégot, 'Critique et interprétation dans les premiers écrits philosophiques d'Adorno', in Theodor W. Adorno, *L'actualité de la philosophie et autres essais*, ed. by Jacques-Olivier Bégot (Paris: Éditions Rue d'Ulm, 2018), pp. 71–103 (p. 85).

of life'.[55] Apart from Horkheimer's theme of injustice and suffering inflicted on animals by humans, discussed in the next paragraph, in this aphorism, Adorno finds in the establishment of zoos the same hope for the advent of a better species. In zoos, even more so when the animals are not in cages but the barriers separating them from the public are invisible (trenches), domination over nature is even more ruthless: 'The more purely nature is preserved and transplanted by civilization, the more implacably it is dominated [*beherrscht*]'.[56] The suffering of animals – more pronounced in cages than in free enclosures – and the progress in terms of humanity found in Hagenbeck's bar-free gardens, express something about 'the inescapability of imprisonment', which is a consequence of history. Indeed, Adorno explains, zoos arose in the 19th century as a product of colonial imperialism: the populations of the areas of Africa and inland Asia colonised by Europeans offered animals as symbolic tributes.[57] The imprisonment of exotic animals in zoos is therefore a consequence of colonial domination and the exploitation of the resources of colonised territories. Adorno seems to suggest a logical and consequential continuity between European imperialist expansion (involving the large-scale domination of 'savage' peoples) and a planetary domination of nature with the possibility of subjugating ever larger natural units, transplanting natural entities from exotic countries to European countries.[58] In conclusion, Adorno expresses the paradox of a nature merely surviving in an irrational culture (which is integrated into the apparatus of domination and the conventions of 'second nature') and a rationalised culture that absorbs nature into itself (to the point of domesticating wild animals),

---

55 Theodor W. Adorno, *Minima moralia. Reflexionen aus dem beschädigten Leben*, in Adorno, *GS* 4 (Frankfurt am Main: Suhrkamp, 2003), § 74, 'Mammut', p. 130; *Minima moralia: Reflections from Damaged Life*, trans. by Edmund Jephcott (London–New York: Verso, 2005), Part one, § 74, 'Mammoth', p. 115.

56 Adorno, *Minima moralia*, § 74, p. 131; *Minima moralia*, Engl. trans., p. 115.

57 Adorno, *Minima moralia*, § 74, p. 131; *Minima moralia*, Engl. trans., p. 116.

58 As Adorno writes: 'immer größere Natureinheiten'. Adorno, *Minima moralia*, § 74, p. 131; *Minima moralia*, Engl. trans., p. 115.

thus eliminating nature (difference) and, with it, the possibility of reconciliation between nature and culture:

> Only in the irrationality of civilization itself, in the nooks and crannies of the cities, to which the walls, towers and bastions of the zoos wedged among them are merely an addition, can nature be conserved. The rationalization of culture, in opening its doors to nature, thereby completely absorbs it, and eliminates with difference the principle of culture, the possibility of reconciliation.[59]

3. *The problem of nature and natural beings in Horkheimer's* Dämmerung

According to Martin Jay, 'The Idea of Natural History' shows Adorno's 'keen awareness of the costs of subjugating objective nature'. He affirms that '[u]nlike Horkheimer in the 1930s, [Adorno] seems to have been sensitive to the risks involved in dominating nature, whether by capitalist or socialist means'.[60] Nevertheless, in

---

59 Adorno, *Minima moralia*, § 74, pp. 131–2; *Minima moralia*, Engl. trans., p. 116.

60 Martin Jay, *Marxism and Totality: The Adventures of a Concept from Lukács to Habermas* (Berkeley, CA: University of California Press, 1984), p. 261. According to Heidrun Hesse, the young Horkheimer believed that a certain degree of control over nature by society would make it possible to put an end to social domination. Moreover, in his writings from the first half of the 1930s, Horkheimer presented nature as a hostile force, a blind and indiscriminate substrate of human activity. Heidrun Hesse, *Vernunft und Selbstbehauptung. Kritische Theorie als Kritik der neuzeitlichen Rationalität* (Frankfurt am Main: Fischer Taschenbuch, 1984), pp. 50–1. Conversely, Gerhard Schwppenhäuser notes that criticism of social domination was already present in the young Horkheimer and that, in the 1940s, he based it on a philosophical critique of domination over nature. Gerhard Schweppenhäuser, *Zur Aktualität von Max Horkheimer. Einführung in sein Werk* (Wiesbaden: Springer, 2023), p. 4. See also Rolf Wiggershaus, *The Frankfurt School: Its History, Theories, and Political Significance*, trans. by Michael Robertson (Cambridge, MA: The MIT Press, 1995), p. 135. Here he highlights Horkheimer's 'conviction, in view of the advanced state of human control over nature, that humanity's demand for happiness has as its goal the

a different way, particularly in *Dämmerung* (but also, in a certain sense, in 'Traditional Theory and Critical Theory'),[61] Horkheimer also reveals a specific sensitivity to the subjugated condition of natural entities, at least as far as Western societies are concerned.

In 1934, when Horkheimer was already exiled, he published under the pseudonym Heinrich Regius a book of aphorisms, written between 1926 and 1931, entitled *Dämmerung: Notizen aus Deutschland*. It offers an examination of bourgeois life that closely reflects the suffering experienced by those who struggle to find their place within the bourgeois lifestyle. Thus, developing a social critique of 'bourgeois society',[62] it reflects in particular on the gap between poverty and wealth. *Dämmerung* depicts the peculiar social situation in Germany in the 1920s and, more generally, the capitalist societies' way of life in the throes of economic crisis. This text begins to reveal Horkheimer's acute sensitivity to contradictions of bourgeois society, between the moral and humanist aspirations of bourgeois ideals in the early modernity and the material realities of

---

reconciliation in practice of particular and general interests on the basis of a planned economy'.

61 See Chapter III, § 3 of this book.

62 On the concept of 'bourgeois society', see Pamela Pilbeam, 'Bourgeois society', in *A Companion to Nineteenth-Century Europe: 1789-1914*, ed. by Stefan Berger (Oxford–Malden, MA: Blackwell Publishing Ltd, 2006), pp. 86–97. Consider that the German term 'bürgerlich' is considered highly problematic by translators. It is an adjective related to the noun 'Bürger', which may be translated as 'bourgeois' or 'citizen'. 'Bürgerlich' retains both connotations. In expressions such as 'civil code', 'civil society', 'bourgeois society', 'bourgeois strata', and 'bourgeois family', the German term for 'civil', 'civic', and 'bourgeois' is 'bürgerlich'. But 'bürgerlich' also designates 'middle class' as opposed to 'noble' or 'peasant'. This meaning also appears in Horkheimer's texts. Here we find also the word 'die Kleinbürger'. See for example a passage to designate the members of the petite bourgeoisie: Max Horkheimer, *Dämmerung. Notizen in Deutschland*, *GS* 2: *Philosophische Frühschriften 1922–1932*, ed. by Gunzelin Schmid Noerr (Frankfurt am Main: Fischer, 2012 [1987]), pp. 309–452 (p. 344); *Dawn: Notes 1926-1931*, in Horkheimer, *Dawn and Decline: Notes 1926-1931 and 1950-1969*, trans. by Michael Shaw with an afterword by Eike Gebhardt (New York: The Seabury Press, 1978), pp. 15–112 (p. 37).

alienation, misery and suffering to which its form of life condemns the majority. The diagnosis and analysis of these contradictions would be the specific work of critical theory, as Horkheimer defines it in the second part of the 1930s.[63] The title *Dämmerung* evokes the catastrophe of a twilight of the gods and alludes to a quote from the German-language Austrian poet Nikolaus Lenau, which Horkheimer uses as the motto for his book and which speaks of 'dying in the twilight'.[64] Horkheimer uses the term metaphorically, ambivalently and vaguely prophetically. Consider for example the last sentence of the first aphorism: 'The enemies of the Inquisition turned that dusk into the dawning of a new day. Nor does the dusk of capitalism have to usher in the night of mankind although today it certainly seems to be threatening it'.[65]

The text offers a significant reflection on nature and the condition of natural beings in capitalistic societies. In an important aphorism Horkheimer illustrates 'today's social structure' through the image of a skyscraper. Starting from the top, that is 'the feuding tycoons of the various capitalist power constellations', and passing through the various minor leaders, functionaries, intellectuals, workers and the poor of Western cities, it arrives at the subjugated peoples in the colonies and, finally, at animals: 'Below the spaces where the coolies of the earth perish by the millions, the indescribable, unimaginable suffering of the animals, the animal hell in human society, would

63 'To put it in broad terms, the theory says that the basic form of the historically given commodity economy on which modern history rests contains in itself the internal and external contradictions of the modern era; it generates these contradictions over and over again in an increasingly heightened form; and after a period of progress, development of human powers, and emancipation of the individual, after an enormous extension of human control over nature, it finally hinders further development and drives humanity into a new barbarism'. Max Horkheimer, 'Traditionelle und kritische Theorie', *GS* 4: *Schriften 1936-1941*, ed. by Alfred Schmidt, 2nd edition (Frankfurt am Main: Fischer, 2009), pp. 162–216 (p. 201); 'Traditional and Critical Theory', in Max Horkheimer, *Critical Theory: Selected Essays*, trans. by Matthew J. O'Connell and others (New York: Continuum, 1972), pp. 188–243 (p. 227). (translation amended).

64 Horkheimer, *Dämmerung*, p. 311.

65 Horkheimer, *Dämmerung*, p. 313; *Dawn: Notes 1926-1931*, p. 17.

have to be depicted, the sweat, blood, despair of the animals'.[66] It should be noted that in *Dialectic of Enlightenment*, Horkheimer and Adorno, speaking about the (Kantian) concept of harmony (*Zusammenstimmung*) between the subject's cognitive faculty and nature, as well as the development of Enlightenment in modern science and industry,[67] would write: 'Being is apprehended in terms of manipulation [*Verarbeitung*] and administration [*Verwaltung*]. Everything – including the individual human being, not to mention the animal – becomes a repeatable, replaceable process, a mere example of the conceptual models of the system'.[68] In the capitalistic system, animal always appears on the bottom floor – if not in the basement – of the skyscraper of domination and administration: it is something we avoid talking about, but its exploitation and suffering are part of man's generalised domination over nature. It is no coincidence that the aphorism 'The Skyscraper' concludes with these images: 'The basement of that house is a slaughterhouse, its roof a cathedral, but from the windows of the upper floors, it affords a really beautiful view of the starry heavens'.[69]

As Horkheimer writes in another aphorism of *Dämmerung*, 'Baseness is rewarded', capitalism in its current phase is 'exploitation organised on a global scale'. The system is built and sustained through the exploitation of people and natural resources on a global scale. Indeed, Horkheimer notes that every thought or action for 'the apparatus of exploitation encompassing the entire world [*für den weltumspannenden Ausbeutungsapparat*]' involves the possibility of affirming oneself.[70] Therefore, this apparatus encompasses human beings, animals and natural resources. In other words, the history

---

66 Horkheimer, *Dämmerung*, 'Der Wolkenkkratzer', p. 380; *Dawn: Notes 1926-1931*, 'The Skyscraper', p. 66.

67 Horkheimer and Adorno, *Dialektik der Aufklärung*, p. 105–7; *Dialectic of Enlightenment*, pp. 64–5.

68 Horkheimer and Adorno, *Dialektik der Aufklärung*, pp. 106–7; *Dialectic of Enlightenment*, p. 65.

69 Horkheimer, *Dämmerung*, 'Der Wolkenkkratzer', p. 380; *Dawn: Notes 1926-1931*, 'The Skyscraper', p. 67.

70 Horkheimer, *Dämmerung*, 'Auf die Gemeinheit ist eine Prämie gesetzt', p. 335.

and destiny of non-human living beings and natural resources is inextricably intertwined with the apparatus of domination set up by capitalist society.

On various occasions, in *Dämmerung*, Horkheimer brings to light the continuity in the suffering that unite and equate some human beings and animals. This generalised, 'shared suffering'[71] is the effect of 'a bad practice' that has emptied of meaning the very concepts with which capitalist ideology covers up its misdeeds, first and foremost the concept of 'humanity'.[72] Consider particularly the aphorism: 'Europe and Christianity', in which Horkheimer reflects on the gulf existing between the moral criteria of Christians and their actual conduct.

He begins by focusing on the unscrupulousness of the ruling classes in Christian countries, who are prepared to consider any abominable action that serves their interests to be lawful. With a not-so-veiled allusion to Machiavelli, Horkheimer notes that – from the extermination of millions of young people in war to the assassination of political opponents – there is no crime that the public conscience cannot in some way be brought to accept.[73]

In the second part of this aphorism Horkheimer starts to reflect on the shared suffering of men and animals:

> That Christians remain unruffled when confronted with the misery of others; that they do nothing to help where injustice is done to the powerless but themselves torture children and animals; that they calmly pass by the walls behind which misery and despair take their course because their interests must be furthered; that it is always a misfortune to fall into their hands; that in view of all this, they daily worship someone as their divine model who, they believe, sacrificed himself for humanity, this lie marks every step of European life.[74]

71 Horkheimer, *Dämmerung*, 'Monadologie', p. 314; *Dawn: Notes 1926-1931*, p. 18.

72 Horkheimer, *Dämmerung*, 'Entehrte Begriffe', p. 315; *Dawn: Notes 1926-1931*, 'Debased Concepts', p. 18.

73 Horkheimer, *Dämmerung*, 'Europa und Christentum', pp. 411–12; *Dawn. Notes 1923-1931*, 'Europe and Christianity', p. 88.

74 Horkheimer, *Dämmerung*, p. 413; *Dawn. Notes 1923-1931*, p. 90.

He then examines cases in which men – even young boys – cause gratuitous suffering to animals: for example, a case he witnessed first-hand involving wealthy, pious people who, in an enchanting mountain setting, enjoyed shooting at pigeons that had been kept in a dark box for some time and were shot as soon as they were released, while dazzled by the sunlight. In the same context, in the evening, a film is shown at the cinema depicting an expedition into the jungle during which a lamb is tied up in a trap to lure a leopard, which kills the lamb and is then killed by humans.[75] Horkheimer comments on these useless sacrifices of animals that are symbolic for the Christian religion itself:

> But the lamb is a religious symbol. Usually, primitives [*Die Wilden*] do not devour their sacred or totem animals. The Christians make symbols of them. They don't worship the animals but in or through them they revere the deity, and this is the reason the animals are not spared in reality. They have not profited from the sublimation of our ideas about the godhead.[76]

It is interesting to reflect on the implications of Horkheimer's considerations: Christianity attributes symbolic value to certain animals but does not seem to care for them in reality; indeed, it admits that they can be killed, even for non-essential purposes, while totemic populations develop a more complex relationship with natural entities: totemic animals are normally neither killed nor eaten.

At various points in the text, tangentially analysing 'the general machinery of exploitation [*Ausbeutungsmaschine*]',[77] Horkheimer evokes the common condition of humans and animals within a chain of suffering.[78] In 'The Impotence of Renunciation', addressing a possible idealist who refuses to come to terms with the compromises and choices of politicians, Horkheimer notes:

---

75 Horkheimer, *Dämmerung*, p. 414; *Dawn. Notes 1923-1931*, pp. 90–1.

76 Horkheimer, *Dämmerung*, p. 414; *Dawn. Notes 1923-1931*, p. 91.

77 Horkheimer, *Dämmerung*, 'Ohnmacht des Verzichts', p. 406; *Dawn. Notes 1923-1931*, 'The Impotence of Renunciation', p. 84.

78 See, for example, Horkheimer, *Dämmerung*, 'Eine Neurose', p. 449; *Dawn. Notes 1923-1931*, 'A Neurosis', p. 110.

> Your refusal to continue to profit from the large-scale torture inflicted on man and beast, your resolve to renounce comfort and security, will spare no human being, no animal, any suffering. Nor should you hope that enough people will imitate you. In recent history, the propaganda of personal renunciation, of individual purity, has always been used by the powerful to keep their victims from doing something more dangerous. It has degenerated into sectarianism. The progressive reduction of misery is the result of long, world-historical struggles whose phases are defined by successful and unsuccessful revolutions. Every success involves the danger of terrible reversals, of new barbarism, of heightened suffering. It is not compassion but intelligence, courage, organizational skill that enable one to participate here. If you lack those qualities, you can do nothing for society.[79]

Immediately afterwards, Horkheimer clarifies that awareness of the futility of individual refusal to consent to the general exploitation of humans and animals certainly does not justify the opposite choice: participating in oppression (*Unterdrückung*).[80] The only possible attitude towards the economic system in which human beings and other living beings become equally subject to manipulation and exploitation, is to join the historic struggles of global significance, with an awareness of their ups and downs; successes may be followed by failures, and therefore new and increased suffering. Therefore, the fight against the misery of human and animal conditions does not require compassion but specific qualities functional to the revolutionary struggle, including foresight, courage, and organizational skills. While in other texts Horkheimer reflects on the complexity and beneficial effects of compassion in light of Schopenhauer's doctrine,[81] but still

79 Horkheimer, *Dämmerung*, 'Ohnmacht des Verzichts', p. 406; *Dawn. Notes 1923-1931*, 'The Impotence of Renunciation', p. 84.

80 Horkheimer, *Dämmerung*, pp. 406–7; *Dawn. Notes 1923-1931*, p. 84.

81 Horkheimer is well aware of the relevance of this Schopenhauerian concept and its theorisation in the *Preisschrift über die Grundlage der Moral*. He cites it, for example, in a university course on Leibniz in 1958: 'In morality, too, I recognise the true relationship between myself and others; the principium individuationis is broken through in compassion, his suffering becomes mine. Then I act rightly in the deepest sense; I do not have myself alone in mind. In the foundation of morality Schopenhauer measures his moral principle against others' (translation by the author). Max Horkheimer,

emphasising its theoretical limitations,[82] in this text,[83] he distances himself from Schopenhauer – one of his reference authors who accompanied his philosophical reflection throughout his entire life[84] –, who elaborates a complex theory of compassion. According to Schopenhauer, compassion, 'the great mystery of ethics', is an anti-egoistic feeling of metaphysical origins that neutralizes egoism as the rule that normally guides the behaviour of humans and nonhuman animals.[85] Moreover, he thinks that true virtue consists in the inclination to feel empathy for the sufferer and to take compassionate actions; it comes from 'intuitive cognition that recognizes in another individual the same essence as in its own'.[86] Not only does

---

Der Begriff der Seele seit Leibniz, *GS* 13: *Nachgelassene Schriften 1949–1972*, ed. by Gunzelin Schmid Noerr (Frankfurt am Main: Fischer, 1989), pp. 516–69 (p. 554). See also Horkheimer, *Späne. Notizen über Gespräche mit Max Horkheimer, in unverbindlicher Formulierung aufgeschrieben von Friedrich Pollock*, in Horkheimer, *GS* 14: *Nachgelassene Schriften 1949–1972 5. Notizen*, ed. by G. Schmid Noerr (Frankfurt am Main: Fischer, 1988), pp. 177–547 (pp. 383, 387, 402, 483).

82 'Schopenhauer fundamentally falls short of this insight. He only sees compassion for the individual and not mediation by the community [Schopenhauer bleibt im Grund hinter dieser Einsicht zurück. Er sieht nur das Mitleid mit den Einzelnen und nicht die Vermittlung durch die Gemeinschaft]'. Horkheimer, *Späne*, 'Individuum und Volk', p. 402 (translation by the author).

83 References to compassion also emerge in the aphorisms 'Humanity' and 'The Freedom of Human Choice': Horkheimer, *Dämmerung*, pp. 400 and 410, respectively; *Dawn. Notes 1923-1931*, pp. 79–80 and 87, respectively.

84 See Rolf Wiggershaus, *Max Horkheimer zur Einführung* (Hamburg: Junius Verlag, 1998), p. 57. Wiggershaus highlights the presence, in the short novels of the young Horkheimer, of an ambivalent assessment of compassion, which is influenced as much by the legacy of Schopenhauer (compassion as an expression of solidarity among all living beings) as by that of Nietzsche (compassion as a means of blackmail, an instrument of power that unmasks the limited, unfree nature of living beings). Ibid., pp. 29–33.

85 See Arthur Schopenhauer, *Prize Essay On the Basis of Morals* [*Über die Grundlage der Moral*] (1840), trans. by Christopher Janaway, in Schopenhauer, *The Two Fundamental Problems of Ethics* (1841/1860) (Cambridge: Cambridge University Press, 2009), pp. 113–258 (p. 201).

86 Arthur Schopenhauer, *The World as Will and Representation* [*Die Welt als Wille und Vorstellung*], vol. 1 (1818/1844/1859), trans. and ed. by

Horkheimer declare the futility of compassion in relation to the misery generated by the general machine of exploitation, but he also believes that only conscious struggle, with all the uncertainties that struggles and revolutions entail, can contribute to improving the conditions of humans and animals. The emphasis on 'consciousness' also constitutes a departure, at least in these pages, from Schopenhauer, for whom compassionate feelings are not always accompanied by consciousness: they can also be a kind of cross-cutting theme that does not require any explicit consciousness.[87] Contempt for injustice and empathy for those who suffer are of little use unless we take concrete action by joining the great struggles against the machinery of exploitation or, as Horkheimer writes in another aphorism, 'the giant machinery [*der Reisenapparat*] for the preservation of injustice and misery [*zur Aufrechterhaltung von Ungerechtigkeit und Elend*]'.[88] This apparatus – that produces 'dumbing down' (*Verdummung*) and 'oppression' (*Unterdrückung*) and perpetuates inequality in the present for the profit of a small minority[89] – also encompasses non-human animals and natural resources. Indeed, the same (capitalist) apparatus producing the misery and suffering of a part of humanity (wage-earning workers, small farmers, unemployed people) for the

Christopher Janaway, Judith Norman, and Alistair Welchman (Cambridge: Cambridge University Press, 2014), p. 394. On this topics, see Sandra Shapshay, 'Schopenhauer on the Moral Considerability of Animals: Toward a Less Anthropocentric Ethics', in *The Palgrave Schopenhauer Handbook*, ed. by Sandra Shapsay (Cham, Switzerland: Palgrave Macmillan, 2017), pp. 283–98; Matthew C. Altman and Cynthia D. Coe, 'Wolves, Dogs, and Moral Geniuses: Anthropocentrism in Schopenhauer and Freud', in *The Palgrave Schopenhauer Handbook*, pp. 447–71.

87 'For awakening compassion, which has been proved as the sole source of disinterested actions and consequently as the true basis of morality, no abstract cognition was required, but only intuitive cognition, the simple grasp of the concrete case, to which compassion responds at once without further mediation of thought'. Arthur Schopenhauer, *Prize Essay On the Basis of Morals*, p. 232.

88 Horkheimer, *Dämmerung*, 'Dienst am Geschäft', p. 398; *Dawn. Notes 1923-1931*, 'In the Service of Business', p. 79.

89 Horkheimer, *Dämmerung*, 'Verpönte Affekte', p. 443; *Dawn. Notes 1923-1931*, 'Forbidden emotions', p. 107 (translation amended).

benefit of a small minority, 'causes the unending fear of death and torture of millions of those animals'.[90]

The idea of a continuity of suffering shared by humans and animals, returns significantly in an aphorism of *Notizen*, dated 1961–62, entitled 'With Open Eyes [*Mit offenen Augen*]', in which Horkheimer encourages a rethinking of the awareness of physical and moral pain, particularly physical torture, which we can find in specific institutions and establishments:

> *With Open Eyes*: To be conscious of the untold, horrible physical and psychological pain, and particularly physical torture, which is suffered at every moment in penitentiaries, hospitals, slaughterhouses, behind walls and in full view the world over, to see all this means to live with open eyes. Without such awareness, every decision is blind, every sure step a misstep, every happiness untrue. But happiness and truth, like truth and grief, are one. This is what Christianity means where it is not betrayed by its mindless adherents.[91]

It should be noted that shared and generalised suffering manifests itself in, and is linked to, specific institutional locations and establishments: as outlined in various aphorisms – with clear references or allusions – penitentiaries, hospitals, slaughterhouses physically embody the structuring of places where humans and animals are tortured and suffer. It is in these institutions and establishments that we find humans and animals in their concrete form, while philosophy (the new philosophical anthropology) and science (physiology, for example) seek and describe the essence of humans – isolated and abstract humans – distinguishing it from

90 Horkheimer, *Dämmerung*, 'Die neue Sachlichkeit', p. 425; *Dawn. Notes 1923-1931*, 'The New Objectivity', pp. 97–8 (translation amended).

91 Max Horkheimer, *Notizen 1949-1969*, in Horkheimer, *GS* 6: *Zur Kritik der instrumentellen Vernunft und Notizen 1949–1969*, ed. by Alfred Schmidt, 2nd edition (Frankfurt am Main: Fischer, 2008), pp. 187–425 (p. 395); *Decline. Notes 1950-1969*, in Horkheimer, *Dawn and Decline: Notes 1926–1931 and 1950–1969*, trans. by Michael Shaw with an afterword by Eike Gebhardt (New York: The Seabury Press, 1978), p. 222.

that of all other living beings.[92] Nevertheless, the new anthropology 'disregards the fact that it is not abstract individuals that kill animals and eat their cadavers and thus cause the unending fear of death and torture of millions of those animals'.[93]

In conclusion to this chapter, it should be noted that the texts examined here show how Adorno and Horkheimer criticise the rigid distinction between nature and history. In this dichotomy it is possible to see the distinction between the first and the second nature, in brief, the antithesis between nature and culture. Therefore, their thesis seems to fuel contemporary reflections that seek to radically rethink the relationship between nature and culture, between human beings and the natural environment. It is interesting to note that today, in a different manner, in *Facing Gaia*, Bruno Latour affirms that we cannot thematise nature and culture as two separate domains:

> We are not dealing with *domains* but rather with one and the same *concept* divided into two parts, which turn out to be bound together, as it were, by a sturdy rubber band. In the Western tradition, we never speak of the one without speaking of the other: there is no other nature but this definition of culture, and no other culture but this definition of nature.[94]

In the second lecture of *Facing Gaia*, Latour compares the behaviour of a literary character (a 'human subject who should be in full control and able to achieve his intentions' but who 'is precisely the one who is made to act by objective forces that he cannot "check"'), two competing rivers (the Mississippi and the Atchafalaya) and their movement, and the factor that releases corticotropin.[95] In this regard, he writes:

---

92 Horkheimer, *Dämmerung*, 'Die neue Sachlichkeit', pp. 424–5; *Dawn. Notes 1923-1931*, 'The New Objectivity', p. 97.

93 Horkheimer, *Dämmerung*, p. 425; *Dawn. Notes 1923-1931*, pp. 97–8.

94 Bruno Latour, *Face à Gaïa. Huit conférences sur le nouveau régime climatique* (Paris: La Découverte Poche, 2023 [2015]), p. 25; *Facing Gaia: Eight Lectures on the New Climatic Regime* (Cambridge: Polity Press, 2017), p. 15.

95 Latour, *Face à Gaïa*, pp. 68–9, 71, 74–5, respectively; *Facing Gaia*, pp. 50–1, 53, 55–6, respectively. Note the reference to 'Marshal Kutuzov's

> In the final analysis, the distinction between humans and nonhumans has no more meaning than the Nature/Culture distinction. It would be just as artificial a distinction as putting Kutuzov and the Army Corps of Engineers in one box and the Mississippi and the CRF in another, as though the first were characterized by a form of soul or consciousness or mind and as though the second were, if not inert, then at least lacking in goals and intentions. The distinction between humans and nonhumans and the difference between culture and nature have to be treated the same way: to be sure that we are not using them as resources but rather as objects of study, we have to go a level deeper, to the common concept that distributes the figures into separate parts. To believe that these terms describe anything at all about the real world amounts to taking an abstraction for a description.
>
> When we claim that there is, on one side, a natural world and, on the other, a human world, we are simply proposing to say, after the fact, that an arbitrary portion of the actors will be *stripped of all action* and that another portion, equally arbitrary, will be *endowed with souls* (or consciousness).[96]

According to Bruno Latour, a subject is not a being 'acting in an autonomous fashion in relation to an objective context'; rather, a subject shares 'agency with other subjects that have also lost their autonomy'.[97] A river or the factor that releases corticotropin have the same titles and rights to be called subjects, beings endowed with 'agency'. In interacting with these subjects, 'or rather quasi-subjects', it is necessary to renounce the claim of control over non-human beings and 'stop fearing the nightmare of finding ourselves once again prisoners of "nature"'. Nonhuman beings are not inert beings compared to which we recognise ourselves as agents but, on the contrary, they reveal themselves as 'agencies that are no longer without connection to what we are and what we do'.[98] In brief, Latour draws our attention to the troubling 'idea of an Earth that would react, retroactively, to our actions'.[99] Latour's position is consistent

(non-)decision on the eve of the famous Battle of Tarutino on October 12, 1812'.

96 Latour, *Face à Gaïa*, p. 78; *Facing Gaia*, p. 58.

97 Latour, *Face à Gaïa*, p. 83; *Facing Gaia*, p. 62.

98 Latour, *Face à Gaïa*, p. 83; *Facing Gaia*, p. 62.

99 Latour, *Face à Gaïa*, p. 85; *Facing Gaia*, p. 63.

with the work of Philippe Descola. Illustrating the Achuar people's view of the relationship between humans and non-humans, who live on both sides of the border between Ecuador and Peru, Descola writes: 'nature is no transcendent element nor simply an object that needs to be socialized. Rather, it is a subject in a social relationship. It is an extension of the world of the homestead, and in truth it is domesticated even in its most inaccessible reaches'.[100]

Once the necessary distinctions have been drawn, this is the same principle of the interconnection between humans and non-humans that can be detected in Horkheimer and Adorno, the principle of the impossibility of separating the world of human beings (and its history) from the world of natural entities, which is also in a state of flux. This question and the possible comparison between critical theory and certain contemporary issues concerning the nature-culture nexus will be explored in depth in the following chapters.

100 Philippe Descola, *Par-delà nature et culture* (Paris: Gallimard, 2005), p. 23; *Beyond Nature and Culture*, trans. by Janet Lloyd (Chicago: The University of Chicago Press, 2013), p. 15.

II.
HUMANISM IN HORKHEIMER'S ESSAYS OF THE SECOND PART OF THE 1930S

1. *The individual and society in the bourgeois order and anthropology*

Horkheimer's project of reconstructing the interrelationships between the historical and socio-economic processes that mark modern times, bourgeois society and the movement towards self-understanding of the modern world through the speculative work of philosophers, scientists, and politicians, developed throughout the 1930s. This chapter will examine Horkheimer's analysis of bourgeois society in modern times, in order to detect his evaluation of the humanism of early modernity.

In 'Remarks on Philosophical Anthropology' (1935) Horkheimer notes that the principle of birth, which marked the Middle Ages, was an expression of pure accidentality; no one was responsible for it. With respect to this conception, the bourgeoisie performs a double shift: it eliminates the idea of chance and proclaims the inherent equality of all men. What now becomes crucial in the existence of man and society is no longer birth but the ability to shape one's own life through work and performance (*Leistung*). *Leistung* is a term that means both performance and productivity as well as work; it opposes the principle of *Geburt* (birth). The former is linked with the principle of self-preservation and property growth. According to Horkheimer, 'the principle of performance' (*das Prinzip der Leistung*) on which modern society is based 'is basically correct [*ist schon richtig*]'. Indeed, during the modern era, characterised by the decline of the idea that power and dignity are based solely on birth, people come to understand that enjoyment depends not on transcendent divine forces but on their own labour. However, a further step is required, namely, to recognise that the concept of the 'individual' cannot be hypostatised and absolutised, since it 'is determined by the dynamic of the entire society'. In other words, everything that an individual achieves is determined by a combination of subjective and objective, individual and collective, cultural and economic factors.[1] These are fundamental themes to which Horkheimer returns on several occasions during the 1930s.

The bourgeois order is based on the radicalisation of the 'possessive instincts of all individuals [*der Eigentumsinstinkte aller Einzelnen*]' and the preservation of the social totality through their tendency to self-preservation and the race to increase their goods.[2] In modern times, the tendency of human beings to adapt to society's living conditions produces a situation in which the entire stake of existence revolves around individual self-interest: all physical and mental energies are directed towards personal advantage, one's own economic gain. This process, on the other hand, is not accompanied by the full consciousness of the individual, by common reflection and decision.

> The life of the general public [*das Leben der Allgemeinheit*] arises blindly, accidentally, and defectively out of the chaotic activity of individuals, industries, and states. This irrationality expresses itself in the suffering of the majority of human beings. The individual, completely absorbed in the concern for himself and 'his own', does not only promote the life of the whole [*das Leben des Ganzen*] without clear consciousness [*ohne kläres Bewußtsein*]; rather, he effects through his labour both the welfare and the misery of others – and it can never

1 Max Horkheimer, 'Bemerkungen zur philosophischen Anthropologie', *GS* 3, pp. 249–76 (pp. 269–70); 'Remarks on Philosophical Anthropology', in Max Horkheimer, *Between Philosophy and Social Science: Selected Early Writings*, trans. by G. Frederick Hunter, Matthew S. Kramer, and John Torpey (Cambridge, MA–London: The MIT Press, 1993), pp. 151–75 (pp. 169–70).

2 Max Horkheimer, 'Materialismus und Moral', *GS* 3, pp. 111–49 (p. 116); 'Materialism and Morality', in *Between Philosophy and Social Science*, pp. 15–47 (p. 19).

> become entirely evident to what extent and for which individuals his labour means the one or the other.[3]

Indeed, as Horkheimer writes in 'Authority and the Family' (1936), a few years after 'Materialism and Morality', about the societies of his time, marked by the consolidation of totalitarian states:

> In the present economic system society appears to be as blind as subrational [*bewußtlose*] nature. For men do not use communal reflection and decision to regulate the process by which they earn their living in association with others. Instead, the production and distribution of all the goods needed for life take place amid countless uncoordinated actions and interactions of individuals and groups.[4]

The material conditions in which individuals work and reproduce common life – guided by the logic of reciprocal satisfaction of needs but without full awareness of the effects of their labour – ultimately distort the relationships they establish with others, causing unnecessary and irrational suffering.[5] While individuals, by liberating themselves from the supra-individual powers and constraints of late medieval society, have developed a certain self-awareness as autonomous beings, this awareness remains abstract. They are unable to fully grasp the connection between

3 Horkheimer, 'Materialismus und Moral', p. 116–17; 'Materialism and Morality', p. 20.

4 Horkheimer, 'Autorität und Familie', *GS* 3, pp. 336 –417 (p. 371); 'Authority and the Family', in Horkheimer, *Critical Theory: Selected Essays* (New York: Continuum, 2002), pp. 47–128 (p. 81).

5 In a society in which the harmony between the universal interest and that of individuals is neither democratic nor planned, but blind and violent, isolated individuals, in order to survive, are forced to conform to the collectivity; thus, many of them are sacrificed in the name of the collectivity. As David Borman notes, Horkheimer's commitment to substantive materialism, sympathetic to this suffering – for entirely historical reasons that find their justification in a non-ideal social theory – recalls explanatory materialism as unmasking the obstacles to human happiness. David A. Borman, 'Materialism in Critical Theory: Marx and the Early Horkheimer', in *The Palgrave Handbook of Critical Theory*, ed. by Michael J. Thompson (New York: Palgrave Macmillan, 2017), pp. 207–29 (p. 223).

their own activity and the social totality. Consequently, it remains completely unclear to them how their work contributes to the life and development of society, which therefore appears as 'a natural occurrence [*Naturgeschehen*]'. In the bourgeois era, individuals who are prisoners of 'a false consciousness [*ein falsches Bewußtsein*]', trapped in a set of beliefs that sanctify the inevitability of existing social relationships, perceive the generality, that is to say, the social whole as a need, an admonition, a moral anxiety weighing upon the individuals throughout their entire existence.[6] In this case too – as in Lukács and Adorno – the false naturalisation of socio-economic relations that are produced by specific factors at work in the historical organisation of human societies can be seen. Not by chance, already in *Dämmerung*, Horkheimer shows that laws of economy are considered as laws of inanimate nature (*der toten Natur*) in a such a way that the bourgeois thinking accepts this reality as 'superhuman' and 'fetishizes the social process' to the point of calling it blind fate (*Schicksal*) or interpreting it in mystical terms.[7] In this way, bourgeois thinking not only surreptitiously represents economic and social phenomena and connections as natural, but also completely obscures nature from its horizon. In short, it prevents dialectical thinking about the connection between natural entities, non-human living beings, and humans who socially modify the natural environment to satisfy their needs.

In this framework, Kantian morality, which relies on good will while abstracting from the meaning and consequences of individual actions in their specific historical contexts, reveals its idealistic character and contributes to nourishing 'the modern mysticism of sacrifice and obedience'.[8] This idealistic trait is reflected in the idea implied by Kant's doctrine, according to which 'all would be right

---

6 Horkheimer, 'Materialismus und Moral', p. 117; 'Materialism and Morality', p. 20.

7 Horkheimer, *Dämmerung*, 'Die Kleine Mann und die Philosophie der Freiheit', p. 361; *Dawn: Notes 1926-1931*, 'The Little Man and the Philosophy of Freedom', p. 50.

8 Horkheimer, 'Materialismus und Moral', p. 121; 'Materialism and Morality', p. 24.

in the world so long as all were right in Spirit'.[9] On the other hand, insofar as Kantian theory is confronted with reality,[10] it demonstrates that the categorical imperative cannot be meaningfully realised in today's society, which must therefore be transformed and superseded. In other words, for the maxim of action to hold as a general law, we need a world in which it is unscathed by perplexities and objections – a society that can rationally determine its own interests, i.e., the interests of all its members. Only under such conditions can individuals find it meaningful to align their lives with this project and to participate in it.[11] According to Horkheimer, therefore, Kant's doctrine expresses 'the contradiction [*der Widerspruch*] which had saddled the bourgeoisie throughout its entire epoch: it created and clung to an order which is in tension with its own concept of reason'.[12]

Horkheimer expounds this thesis also in the coeval article 'Materialism and Metaphysics'. In it, he observes that throughout history both dominant and dominated classes presented their claims as universally valid postulates. However, the dominated, in particular, rather grasped the contrast existing between the principles posited as universal by the dominant and the actual conditions of reality. As an example of this is the Peasants' War of 1524–25, where the peasants demanded the concrete application of Christian principles, which entailed a modification of the content of Christianity as understood at the time. The same demand to implement the bourgeois idea of justice in a generalised manner leads to criticising and eventually overcoming the free trade society,

---

9 Horkheimer, 'Materialismus und Moral', p. 122; 'Materialism and Morality', p. 24.

10 Horkheimer examines the four examples Kant uses in *Groundwork of the Metaphysics of Morals* to illustrate moral action, arguing that they reveal the inadequacy of the good will. See Horkheimer, 'Materialismus und Moral', pp. 119–21; 'Materialism and Morality', pp. 22–4.

11 Horkheimer, 'Materialismus und Moral', p. 122; 'Materialism and Morality', p. 25.

12 Horkheimer, 'Materialismus und Moral', p. 123; 'Materialism and Morality', p. 25.

from which the idea originates.[13] These demands manifest as an immanent critique of a specific social order. In order to criticise the bourgeois free market as a violation of the bourgeois society's own alleged idea of justice, one appeals to its crystallised ideals, but reinterprets their meaning. However, this is not merely a matter of criticising the bourgeois order from within by denouncing the betrayal of its original ideals. Rather, it involves dialectically grasping the one-sidedness of a certain representation of a conception of justice connected to the freedom of all individuals to realise their own life project, formulated in the light of bourgeois society's emphasis on free trade. This framework, in reality, does not foster the conditions necessary for the authentic implementation of freedom as a condition for a justice that addresses the needs and interests of all. Therefore, the task is to trace the genealogy of the idea of justice in the bourgeois era, demystifying its claim to trans-historical validity, and showing how it is the historical product of human beings operating within the framework of a particular, class-divided, society. Indeed, Horkheimer writes:

> The proof of the contradiction [*Der Nachweis des Widerspruchs*] between the principle on which bourgeois society was founded and the actual reality [*ihrem Dasein*] of that society involves bringing out how justice was one-sidedly defined in terms of freedom and freedom in terms of negation, and substituting a positive conception of justice by offering a groundplan for a reasonable society. When the concept of justice thus changes its meaning, we glimpse the historical origins of what was originally proclaimed as an eternal principle, and we understand that that concept was an idea proposed by definite individuals and conditioned by relationships within a class society.[14]

Evoking Marxian theory according to which societies are intrinsically contradictory, Horkheimer shows that social contradiction lies within the very framework of a society, within

13 Horkheimer, 'Materialismus und Metaphysik', pp. 81–2; 'Materialism and Metaphysics', pp. 21–2.

14 Horkheimer, 'Materialismus und Metaphysik', p. 82; 'Materialism and Metaphysics', p. 22 (translation amended). See also Horkheimer, 'Materialismus und Moral', p. 137; 'Materialism and Morality', p. 37.

'a self-generating "nonidentity" intrinsic to its structures of social relations', which are therefore not rooted in a stable structure, a stable unitary whole.[15] Indeed, in these various passages from 1930s Horkheimer articles, we can see that the immanent critique does not judge critically what 'is' from a conceptual position outside of its object, a critical standpoint extraneous thereto (such as transcendent values). Moreover, it must be able to locate that 'ought' which is opposed to 'is' as an element of its own context, as a possibility immanent to the existent society. Therefore, the immanent critique questions bourgeois society from its own point of view, while simultaneously criticising this position from the perspective of its historical context.[16] From this viewpoint, building on some of Rahel Jaeggi's reflections, Franck Fischbach has emphasised that immanent critique reveals a cognitive aspect, insofar as it stimulates an understanding of how certain norms and ideals inform societal practices, and is oriented towards the internal contradiction between reality and the norms that shape it. And it is precisely in Horkheimer that the diagnosis and experience of contradiction are clearly articulated to the critical attitude.[17] In the early 1930s, Horkheimer argues that it is wrong to assert that the bourgeois principles of freedom, equality and justice have proved ineffective. Rather, it is the concrete conditions, which do not correspond to those principles, that are untenable. Dialectical critique must precisely demonstrate that the ideas of the Enlightenment and the French Revolution – foundational to a rational society – retain their validity despite changed concrete

15 Moishe Postone, *Time, Labor, and Social Domination: A Reinterpretation of Marx's Critical Theory* (Cambridge and New York: Cambridge University Press, 2003), p. 88.

16 On these general characters of 'immanent critique' see Postone, *Time, Labor, and Social Domination*, pp. 87–8; Robert J. Antonio, 'Immanent Critique as the Core of Critical Theory: Its Origins and Developments in Hegel, Marx and Contemporary Thought', *The British Journal of Sociology*, 32:3 (1981), 330–45 (p. 338).

17 See Franck Fischbach, *Pour la Théorie critique. Raison, nature et société* (Paris: Vrin, 2024), pp. 99–100.

situations.[18] By advocating for the possibility of freedom and emancipation for all human beings, Horkheimer develops his own immanent critique that, by revealing the contradictions between ideal claims and actual context, aims to transform a contradictory and disintegrating situation into something new. As he states in 'Materialism and Morality', the conflicts between the different interests of individuals in the current configuration of bourgeois society – where the contrast between human misery and the potential for improvement in living conditions has never been more strident[19] – stem from the divergence of property relations and the competing interests of conflicting economic powers:

> Only after this antagonistic economic form, whose introduction once meant tremendous progress (including among other things the developmental possibility of self-reliant human beings), has come to be superseded [*abgelöst*] by a social life form [*von einer Lebensform der Gesellschaft*] in which productive property is administered in the general interest not just out of 'good intentions' but with rational necessity, only then will the concordance of individual ends cease appearing to be a miracle. Moreover, individuals at that point cease to be merely the exponents of private ends. Each is no longer simply a monad, but rather, in Kant's language, a 'limb' of society at large [*ein 'Glied' der Allgemeinheit*].[20]

It is opportune here to turn to some texts written by Horkheimer in the second half of the 1930s, where he delves further into the contradictions of bourgeois society by reconstructing the genealogy of modern times as opposed to the medieval world.

In 'Remarks on Philosophical Anthropology' (1935), Horkheimer notes that the principle of birth, which marked the Middle Ages, was an expression of pure accidentality; no one was responsible for it. With respect to this conception, the bourgeoisie performs a double

18 Horkheimer, 'Materialismus und Moral', p. 137; 'Materialism and Morality', p. 37.

19 Horkheimer, 'Materialismus und Moral', p. 135; 'Materialism and Morality', p. 35.

20 Horkheimer, 'Materialismus und Moral', p. 126; 'Materialism and Morality', p. 28.

shift: it eliminates the idea of chance and proclaims the inherent equality of all men. What now becomes crucial in the existence of man and society is no longer birth but the ability to shape one's own life through work and performance (*Leistung*). The problem – and this is another aspect of the notorious great contradiction of bourgeois society – is that, in the bourgeois era, accidentality re-emerges in subtle ways, as the conditions of work and leisure are not the same for everyone. As Erich Fromm would also highlight in *Escape from Freedom* (1941), in his meticulous analysis of modern man's freedom and individualism,[21] those conditions vary according to class.[22] Nevertheless, Horkheimer expresses the hope that simply entering the world may mean happiness again, not through power over others, but through humanity's dominion over nature: 'We have now reached the point at which mere admission into the world means luck, not through the power over others, but rather through the dominion of man over nature. The man who is noble by birth appears again when his opposite, the man who is equal by nature, is transformed from an ideology into a truth'.[23] These considerations are also present – with a slightly different nuance – in the 1935 notes transcribed by Pollock:

21 'I should like to emphasize here that the whole process had a different meaning for the small group of wealthy and prosperous capitalists on the one hand, and on the other hand for the masses of peasants and especially for the urban middle class for which this new development meant to some extent wealth and chances for individual initiative, but essentially a threat to its traditional way of life'. Erich Fromm, *Escape from Freedom*, with a new foreword by S. Sloan (New York and Tokyo: Ishi Press, 2011), p. 44. See also pp. 59–60.

22 Horkheimer, 'Bemerkungen zur philosophischen Anthropologie', p. 271; 'Remarks on Philosophical Anthropology', pp. 170–1.

23 'Jetzt ist es an der Zeit, daß wieder der bloße Eintritt in die Welt das Glück bedeutet, aber nicht durch Macht über andere, sondern durch die Herrschaft der Menschheit über die Natur. Der Adel von Geburt wird wieder eingeführt, wenn sein Gegenteil, die Gleichheit der Menschen, aus einer Ideologie zur Wahrheit wird'. Horkheimer, 'Bemerkungen zur philosophischen Anthropologie', p. 271; 'Remarks on Philosophical Anthropology', p. 171.

> Now is the time for mere entry into the world to mean luck again, but not through power over others, but through the domination of classless humanity over nature. The nobility of birth will be reintroduced when its opposite, the equality of man, becomes truth from a lie. Man changes in history'.[24]

It is interesting to note that 'die Herrschaft der Menschheit über die Natur' appears in both texts, but in the second Horkheimer adds the adjective 'classless' (*klassenlosen*) to the noun 'humanity' (*Menschheit*). In summary, he evokes a future condition in which class divisions will be overcome and dominion over nature will mean its transformation for the equitable satisfaction of everyone's needs. In both texts, however, the analysis reveals, in the face of the contradictions of bourgeois society, a desire for the reintroduction of the principle of birth as it should be, i.e., that anyone coming into the world has a right to happiness. This would involve turning the equality of human beings, to which modern bourgeois society so often pays lip-service, into a reality, rather than a mere lie. In any case, the project of the domination of nature would be a *leitmotiv* and a critical target in *Dialectic of Enlightenment*, which will be examined in the next chapter.

As Horkheimer explains in 'Authority and the Family', the principle of individual performance in theory and practice, proclaimed by the rising bourgeois spirit against the principle of authority acquired solely by descent, finds its philosophical and abstract expression in a fundamental concept of thought, that of the individual. This concept was first clearly formulated by Leibniz: 'a self-enclosed, metaphysical centre of power [*ein in sich abgeschlossenes … metaphysisches Kraftzentrum*], separated from the rest of the world, an absolutely isolated monad which is made self-dependent by God.

---

24 'Jetzt ist es an der Zeit, daß wieder der bloße Eintritt in die Welt das Glück bedeutet, aber nicht durch Macht über andere, sondern durch die Herrschaft der klassenlosen Menschheit über die Natur. Der Adel von Geburt wird wieder eingeführt, wenn sein Gegenteil, die Gleichheit der Menschen, aus einer Lüge zur Wahrheit wird. Der Mensch verändert sich in der Geschichte'. Max Horkheimer, 'Notizen 1935', in Horkheimer, *GS* 12: *Nachgelassene Schriften 1931-1949*, ed. by Gunzelin Schmid Noerr (Frankfurt am Main: Fischer, 1985), pp. 225–49 (p. 246). Translation by the author.

Its destiny, according to Leibniz, lies within its own determination, and its stages of development, its happiness or unhappiness, depend on its own internal dynamism'.[25] Under this conception, the modern category of the individual operates a radical twofold separation of the individual itself from society and nature – a distinction that is linked to other well-known philosophical dualities (thought and being, substance and appearance, mind and body, sensibility and intellect, and so on). In this way, the concept of the free individual forged by bourgeois thought, as opposed to the vision of medieval man, is firmly formulated on a metaphysical basis.[26] However, the idea of the individual left to himself, emancipated from the old authorities, conceals the fact that, in practice, for most concrete individuals this liberation crumbles into subjection to 'fearful exploitation of the factory system', and has to conform to an external power (the market) and to situations that, from time to time, present themselves as pure facts, which do not convey any meaning but are simply to be accepted.[27] Horkheimer has by now diagnosed the fundamental pathology and profound laceration of bourgeois society: on the one hand, on the level of philosophical abstraction, the sovereignty and self-determination of the individual is proclaimed; on the other hand, however, the individual comes up against something extraneous, over which he exercises no sovereignty and which ultimately limits his self-determination.[28] As Horkheimer emphasised since his lecture course of 1927, bourgeois thought takes shape in antithesis to the authority of tradition, contrasting the latter with individual reason as the sole valid source of law and truth.[29] However, the proud claim to recognise no authority that cannot be justified in the presence of

25 Horkheimer, 'Autorität und Familie', p. 367; 'Authority and the Family', p. 77.

26 Horkheimer, 'Autorität und Familie', p. 367; 'Authority and the Family', pp. 77–8.

27 Horkheimer, 'Autorität und Familie', pp. 367–8; 'Authority and the Family', p. 78.

28 Horkheimer, 'Autorität und Familie', p. 368; 'Authority and the Family', p. 78.

29 Horkheimer, 'Autorität und Familie', p. 362; 'Authority and the Family', p. 72.

reason reveals its fragility when confronted with the class-divided society and the extraneous powers that burden the individual's consciousness, which – in the course of early modernity – are not yet understood as the result of human labour.[30]

In 'Egoism and Freedom Movement' (1936), Horkheimer argues that bourgeois thought has produced two major views of human nature: according to the first model, human nature is evil and dangerous (Machiavelli, Hobbes), while according to the second, the individual is naturally good and capable of building harmonious social relationships (More, Rousseau). Nevertheless, these two theories have one thing in common: both condemn and banish human impulses that do not have an immediate social impact and are not explicitly oriented towards harmony and love. On the other hand, the same bourgeois society in whose bosom the new anthropology is elaborated – and which stigmatises self-centredness and pleasure in general – creates and entrenches extremely painful conditions of existence to the extent that it asserts its hegemony.[31] The dominant feelings, then, become indifference and mutual hostility among individuals, families, classes, peoples.[32] All those immersed in the world of free competition, marked by ever more evident, stark economic and social contrasts, and by 'a permanent state of war, both internally and externally [*dauernden Kriegszustands nach innen und außen*]', 'develop the egoistic, exclusionary, hostile sides of their being in order to survive in this harsh reality'.[33] Bourgeois society is not founded on collaborative relations that aim at the happiness of its members, but on the individual and isolated action of a human being who struggles for self-preservation. In the absence of a project

30 Horkheimer, 'Autorität und Familie', p. 369; 'Authority and the Family', p. 79.

31 Horkheimer, 'Egoismus und Freiheitsbewegung. Zur Anthropologie des bürgerlichen Zeitalters' (1936), *GS* 4, pp. 9–88 (pp. 9–13); 'Egoism and Freedom Movement: On the Anthropology of the Bourgeois Era', in *Between Philosophy and Social Science*, pp. 49–110 (pp. 49–52).

32 Horkheimer, 'Egoismus und Freiheitsbewegung', p. 13; 'Egoism and Freedom Movement', p. 52.

33 Horkheimer, 'Egoismus und Freiheitsbewegung', p. 13; 'Egoism and Freedom Movement', p. 52 (translation amended).

capable of identifying a way to satisfy communal needs, individuals believe that they are working for themselves, and that they must only be concerned with their personal survival.[34] This vision of individual starts with a general theory according to which 'for everything in nature, and thus for the body and its indwelling soul, to perish represents the greatest evil, while self-preservation and all actions toward that end constitute the highest good.'[35] In this way, a conception of non-human nature seems to be projected back onto human beings. In reality, the idea that self-preservation is the law and standard of each natural being, corresponds 'to the social existence of the bourgeois individual'.[36] Even in this case, Horkheimer observes an undue shift from the attitude that characterises individuals in bourgeois society to the identification of a norm and a law that would regulate the whole of nature. Surreptitiously, bourgeois ideology always seeks to justify as natural and eternal something – a tendency, a behaviour, a set of practices and conventions – that characterises individual historical epochs and human cultural contexts.

The fundamental contradiction of bourgeois society emerges here: it favours egoism and individualism in the economic and social sphere, while condemning them in its theoretical-spiritual dimension. In this process, the individual comes to know themselves as the isolated subject of economic interests, interacting with other individuals through the market, i.e. through buying and selling. As a result, the condition of alienation, the 'otherness [*Fremdheit*]', rises to an anthropological category.[37] Thus, the typical form of existence

34 Horkheimer, 'Egoismus und Freiheitsbewegung', p. 70; 'Egoism and Freedom Movement', p. 95.

35 Horkheimer, 'Egoismus und Freiheitsbewegung', p. 10; 'Egoism and Freedom Movement', p. 50.

36 Horkheimer, 'Egoismus und Freiheitsbewegung', p. 11; 'Egoism and Freedom Movement', p. 50.

37 Horkheimer, 'Egoismus und Freiheitsbewegung', p. 70; 'Egoism and Freedom Movement', pp. 95–6. This condition of otherness, as Marx and Engels pointed out, concerns both the upper class and the proletariat, but is experienced by them differently: 'The propertied class and the class of the proletariat present the same human self-estrangement [*dieselbe menschliche Selbstentfremdung*]. But the former class feels at ease and strengthened in this self-estrangement, it recognises estrangement as its own power and has

of the bourgeois individual is that of the monad closed upon itself 'in transcendental loneliness'.[38] The notion of 'monad', conceived of as an isolated spiritual substance closed off from the outside world, plays a decisive role in determining the anthropology of modern times. Such a notion, as Horkheimer notes in an earlier article in 1934, perfectly expresses 'the one-sidedness of human beings in the bourgeois epoch', but it overlooks 'the individual's dependence upon the total social life process'.[39]

Interestingly, within this general framework of bourgeois society, even humanism, connected (like the Protestant Reformation) with profound social changes that mark modernity,[40] presents itself, according to Horkheimer, as a Janus-faced phenomenon.

> The humanism [*Der Humanismus*] that pervades the history of the new spirit shows a double face. Directly, it signifies the glorification of the human being as the creator of their own destiny. Human dignity

---

in it the semblance of a human existence. The latter feels annihilated in estrangement; it sees in it its own powerlessness and the reality of an inhuman existence'. Karl Marx and Friedrich Engels, *Die Heilige Familie oder Kritik der kritischen Kritik Gegen Bruno Bauer und Konsorten* (1845), in Marx and Engels, *Werke*, hrsg. vom Institut für Marxismus-Leninismus, beim Zentral-Komitee der sozialistischen Einheitspartei Deutschlands, Vol. 2 (Berlin: Dietz Verlag, 1962), pp. 3–223 (p. 37); *The Holy Family, or Critique of Critical Criticism. Against Bruno Bauer and Company*, in Marx and Engels, *Collected Works*, 12 vols, Vol. 4, 1844–45 (London: Lawrence & Wishart, 2010), pp. 5–211 (p. 36).

38 Horkheimer, 'Egoismus und Freiheitsbewegung', p. 70; 'Egoism and Freedom Movement', p. 96.

39 Max Horkheimer, 'Zum Rationalismusstreit in der gegenwärtigen Philosophie' (1934), *GS* 3, pp. 163–220 (p. 219); 'The Rationalism Debate in Contemporary Philosophy', in *Between Philosophy and Social Science*, pp. 217–64 (p. 263).

40 Horkheimer, 'Egoismus und Freiheitsbewegung', p. 44; 'Egoism and Freedom Movement', p. 76. A preacher such as Savonarola, in the midst of the humanistic era, can be regarded as a forerunner of the reformers: speaking from the pulpit, he enjoins on the people, along with the fear of God, the practice of bourgeois virtues, respect for laws, the pursuit of peace, dedication to work, submission to authority and sacrifice for the nation. Horkheimer, 'Egoismus und Freiheitsbewegung', pp. 45–6; 'Egoism and Freedom Movement', pp. 76–7.

> [*Die Würde des Menschen*] lies in the power to determine oneself independently of the powers [*von Mächten*] of blind nature within and without; it lies in one's power to act [*sie liegt in seiner Macht zu handeln*]. In the society in which this humanism spread, however, the power of self-determination [*die Macht der Selbstbestimmung*] is unevenly distributed; for inner energies depend no less on external destiny than it does on them. The more remote the abstract concept of the human being, as glorified by humanism, was from their real situation, the more pitiful the individuals of the masses had to appear to themselves, and the more the idealistic divinization of Man – as manifested in the concepts of the greatness, genius, and grace-endowed personality of the leader, etc. – resulted in the self-abasement and the self-contempt of the concrete individual. [...] The Reformation, with its morally depressing antihuman pathos, its hatred for the earthworm's vanity, its dark doctrine of predestination, is not so much the opponent of bourgeois humanism as its other, its misanthropic side. It is humanism for the masses, while humanism itself is the Reformation for the wealthy.[41]

Here, the focus shifts from the humanistic ideal of the *homo faber*, who shapes their own destiny and is able to free themselves from the shackles of 'blind nature', to the social and economic conditions of the society in which reflection on human dignity spreads. In the society that sees the flowering of humanism, there is an unequal distribution of the power to self-determine and emancipate oneself from natural constraints – in which, ultimately, the creative determination of one's own destiny consists. In continuity with these reflections and illustrating Marx's position, in his 1957 outline on the history of materialism, Horkheimer spoke of 'eine Spaltung in der Realität', a split in the reality produced by the bourgeois order, since some individuals have a significantly larger portion of goods than others and greater power to control the economic arrangement and society.[42] As Marcuse points out in an essay that appeared in the Institute's journal in the same period, while the bourgeois liberation of the individual carries with it the promise of a new happiness,

41 Horkheimer, 'Egoismus und Freiheitsbewegung', pp. 73–4; 'Egoism and Freedom Movement', pp. 98–9 (translation amended).

42 Horkheimer, *Geschichte des Materialismus (ausgewählte Kapitel)*, *GS* 13, pp. 397–451 (p. 441).

in fact only a minority possesses the tools, i.e. the necessary purchasing power, to secure a happy existence. Consequently, the abstract equality of individuals in the capitalist system is reversed into a concrete inequality[43] and *humanitas* is purely interiorised: 'Humanity becomes an inner state [*Humanität wird zu einem inneren Zustand*]'.[44]

In his 1936 essay 'Egoism and Freedom Movement', Horkheimer actually delves into a problem already addressed in 'Authority and the Family', published in the same year as a general part of the collection *Studies on Authority and the Family*, and edited by Horkheimer himself.[45] He examines 'the contradiction between the existence of the masses, who were not indeed serfs but were exploited in the most terrible way, and that doctrine of man's freedom and dignity which was dominant in philosophy ever since Pico della Mirandola's time'.[46] Humanism reveals itself as a field of tensions in which the praise of *dignitas hominis* conceals a division

43 Herbert Marcuse, 'Über den affirmativen Charakter der Kultur' (1937), in Marcuse, *Kultur und Gesellschaft*, Vol. 1 (Frankfurt am Main: Suhrkamp, 1965), pp. 56–102 (p. 65); 'The affirmative character of culture', in Marcuse, *Negations: Essays in Critical Theory*, with translations from the German by J. J. Shapiro (London: MayFlyBooks, 2009), pp. 65–98 (pp. 71–2).

44 Marcuse, 'Über den affirmativen Charakter der Kultur', p. 71; 'The affirmative character of culture', p. 76.

45 *Studien über Autorität und Familie*, Forschungsberichte aus dem Institut für Sozialforschung, Vorwort von Max Horkheimer (Paris: Alcan, 1936).

46 Horkheimer, 'Autorität und Familie', p. 375; 'Authority and the Family', p. 85. Horkheimer refers to this 'Widerspruch' in a passage in which he thematises the conditions of workers in the eighteenth century and evokes Voltaire's reflection on work that, by necessity, becomes a scourge and on the masquerade of authority before the worker. See Horkheimer, 'Autorität und Familie', pp. 374–5; 'Authority and the Family', pp. 84–5. One can grasp in this dichotomy between the ideal image of man and the real man, the ideological concealment operated by culture, which constitutes a central theme in the investigation of Horkheimer and Adorno. See, for example, Adorno, *Minima moralia*, p. 48; *Minima moralia*, English trans., § 22, p. 43: 'culture creates the illusion of a society worthy of man which does not exist; that it conceals the material conditions upon which all human works rise, and that, comforting and lulling, it serves to keep alive the bad economic determination of existence. This is the notion of culture

and separation between the idealised and transfigured concept of man (a reflection of the wealthy condition of the humanist) and the concrete man of the masses. Unlike those who can retreat into their own interiority, like Montaigne, in the sixteenth century the majority of individuals are forced into drudgery in manufacturing[47] and, during the modern age, they become the cogs in a machine that aims to get those who tend to avoid work used to working.[48] In this way, Horkheimer highlights the bourgeoisie's tendency to focus exclusively on its own interests and its inability to create the conditions for the emancipation of individuals.[49] At the same time, in his essays from the 1930s, Horkheimer presents the appearance of the masses on the scene of European society (and thus the indispensability of the individuals comprising those masses, as wage labourers in commerce, in manufacturing and as soldiers) along with the concomitant (bourgeois) need to contain their aspirations and desires as two essential phenomena that mark modernity.

Horkheimer then highlights the chiastic relationship that exists between humanism and the Reformation. The Reformation movement, in a kind of 'dialectic of humanism,' reveals the anti-humanist side of humanism. Indeed, at the dawn of modernity, humanism is built on a dual polarisation between the abstract concept of man and the concrete man, and between the exceptional individual (the genius, the leader) and the individual of the masses (the wretch).

---

as ideology, which appears at first sight common to both the bourgeois doctrine of violence and its adversary, both to Nietzsche and to Marx'.

47 Horkheimer, 'Montaigne und die Funktion der Skepsis', *GS* 4, pp. 236–94 (p. 246); 'Montaigne and the Function of Skepticism', in *Between Philosophy and Social Science*, pp. 265–311 (p. 273).

48 Horkheimer, 'Autorität und Familie', p. 374; 'Authority and the Family', p. 84.

49 See what Horkheimer writes in the late 1950s in the aphorism 'Functions and Limits of Bourgeois Culture': 'The less the bourgeoisie succeeded in validating the political emancipation of individuals by subjecting society to their conscious, common activity during its time of dominance in the West, the more compulsively it clung to the ideology of the freedom of the individual. The dogma of the inviolable sovereignty of the nation is part of this'. Horkheimer, *Notizen 1949-1969*, p. 310; *Decline: Notes 1950-1969*, 'Function and Limits of Bourgeois Culture', p. 171.

By insisting on one of the two poles, the reformers enjoin individuals to become aware of their sinful and petty condition, to renounce their instincts and internalise self-discipline and dedication to their work. Humanism, on the other hand, represents the Reformation for the privileged few, those who are not forced to do arduous work and who do not have a wretched self-image. The Reformation can proceed in this direction insofar as its pioneers enunciated the principle that 'salvation does not depend on the sacramental performances of a priestly caste, but on the attitude of the individual's soul'.[50] The reformers establish the autonomy of the individual on an ideological level, an autonomy that corresponds to the needs of a changing reality: 'In light of the new economic tasks, the bourgeoisie had to raise its members to a completely different level of self-discipline, responsibility, and zeal for work than they were accustomed to in the old times of a relatively undynamic economy operating according to fixed rules'.[51] And it goes without saying that one can trace back to the reform movement itself the idea of economic success being a sign of, and a reward for moral righteousness, with poverty appearing as proof of the guilt of the individual, capable of arousing among those belonging to the masses a sense of total worthlessness.[52]

---

50 Horkheimer, 'Egoismus und Freiheitsbewegung', p. 53; 'Egoism and Freedom Movement', p. 82.

51 Horkheimer, 'Egoismus und Freiheitsbewegung', pp. 53–4; 'Egoism and Freedom Movement', p. 83.

52 Horkheimer, 'Egoismus und Freiheitsbewegung', p. 73; 'Egoism and Freedom Movement', p. 98. According to Weber, it is thanks to the growth of Calvinism, in particular in the English and Dutch Puritan tradition, that 'intense worldly activity' comes to be seen as the most significant means of being secure in one's election: it is the condition of 'the self-confident saints' that we can find 'in the hard Puritan merchants of the heroic age of capitalism'. Weber, *The Protestant Ethic and the Spirit of Capitalism*, p. 67. In other words, as Weber points out in a footnote, zeal and activism encouraged in the Puritan sphere are ultimately what makes it possible to overcome 'one's own feeling of moral inferiority'. Max Weber, *Die protestantische Ethik und der Geist des Kapitalismus* (1904–1905), in Weber, *Gesammelte Aufsätze zur Religionssoziologie* (Tübingen: J. C. B. Mohr, 1988), vol. 1, pp. 17–206; *The Protestant Ethic and the Spirit of Capitalism*, trans. by Talcott Parsons, introduction by Anthony Giddens (London and New York: Routledge, 2005), p. 187, n. 47.

2. *Humanism in Horkheimer's essays of the 1930s: Critical theory and active humanism*

This paragraph will examine the essay published in 1938 in the *Zeitschrift für Sozialforschung*, entitled 'Montaigne and the Function of Skepticism'. Here, Horkheimer contrasts the scepticism of early modernity with that of his contemporaries, explaining that the former contributed to the destruction of the feudal order and the scholastic value system, unlike the latter, which has a purely conservative character and legitimises the existing order. It is within this historical-theoretical framework – in which neo-humanism also finds its place as a theoretical and political alternative to all those who are dissatisfied with fascism[53] – and through the cultural and political involution of European society in the 1930s, that Horkheimer develops a comprehensive reflection on Montaigne and the distinctive features of his humanism.

In his essay, Horkheimer stresses, first and foremost, the economic, social and political troubles and crises of sixteenth century France. This scenario includes the fundamental coordinates in which the writing of the *Essays* is framed and to which this founding text of modernity constitutes a mirror and a response. With concise and effective touches, Horkheimer sketches a picture of the economic, social and political phenomena that marked Montaigne's era: the expansion of commerce and consequent increase in the means of circulation; the growth of inflation that undermined the ancient nobility; currency devaluation and consequent decrease in workers' wages; the process of large-scale impoverishment of the lower social strata; the migration of peasants from the countryside to the cities, perceived with increasing distress by urban populations, already struggling with progressive internal impoverishment caused by unemployment; escalating misery, multiplication of epidemics and famines triggering popular uprisings in the cities; a general increase

53 Horkheimer examines the neo-humanism of his time in an article from 1938, published alongside the essay on Montaigne, in which he discusses Siegfried Marck's book, *Der Neuhumanismus als politische Philosophie* (Zürich: Aufbruch, 1938). See Horkheimer, 'Die Philosophie der absoluten Konzentration', *GS* 4, pp. 295–307.

in insecurity and deteriorating trade due to poor infrastructure, banditry, and excessive tolls to be paid in the land and river transit areas; the juxtaposition of class struggles between the bourgeoisie and the proletariat, as well as clashes between dominant social groups. The religious wars that are woven into this network of socio-economic processes can be traced back mainly to the economic contrasts between the stronger social strata: it is no coincidence that the warring parties mobilise the masses, taking advantage of their condition of misery to suit their own purposes.[54]

Having identified the fundamental economic and social processes at work in the sixteenth century, Horkheimer highlights the most original and significant aspects of Montaigne's practical orientations and the theoretical stance that he takes in the *Essays*. His analysis develops in a spiral movement around a number of fundamental thematic cores, which are always dissected in close connection with the economic, social, and political context at different points in the essay. Such perspectival lines of Horkheimer's essay on Montaigne bring to light some fundamental problems and nexuses of modern culture and philosophy, in particular: ataraxia, the isolated self and subjectivism; the nexus between scepticism and the Reformation; the relationship between scepticism and humanism.

Horkheimer first shows that Montaigne, confronted with the crisis of his epoch and 'the horrors of the transition to modernity', just like the ancient sceptics facing the decadence of their society, rejects the mirage of absolute certainty.[55] An attitude that is not renunciatory, but rather an expression of a conscious realism, which reflects a scepticism that rests on two fundamental considerations: human reason cannot attain absolute certainties, and the claim to such certainties has, in any case, nefarious consequences for social coexistence. Thus, according to Horkheimer, by going down the path of scepticism, Montaigne's thought leads to religious tolerance

54 Horkheimer, 'Montaigne und die Funktion der Skepsis', pp. 238–9; 'Montaigne and the Function of Skepticism', pp. 267–8.

55 Horkheimer, 'Montaigne und die Funktion der Skepsis', p. 239; 'Montaigne and the Function of Skepticism', p. 268.

and a kind of political conservatism.[56] 'Montaigne also proves – Horkheimer writes – to be a rejuvenator of the old scepticism through his knowledge of the world and his statesmanlike abilities'.[57] The revival of ancient scepticism in the sixteenth century is intimately connected with the biography of Michel de Montaigne, who is by character a diplomat, belonging to the new nobility rooted in the bourgeois soil and who chose essentially the path of negotiation, not of struggle, which allowed him to relate casually and devoutly as much to the monarchy as to the church and, in Germany, to argue passionately and without bias with both Catholic and Protestant clergymen and teachers.[58] This attitude reveals a remarkable breadth of outlook: 'As with the ancients, philosophical scepticism presupposes a broad horizon. It is the opposite of narrow-mindedness. Its style is that of description, not theory'.[59]

According to Horkheimer, while scepticism presupposes a remarkable breadth of horizons – an ancient attitude that Montaigne inherits and renews – it also reveals an indisputable conservative approach. Human beings do not direct their actions in the light of absolute criteria and cognitions (which, in any case, do not exist), but mainly on the basis of prejudices and habits. Precisely because, in the gnoseological sphere, it cannot be established that one opinion should be favoured over another, in the concrete sphere the sceptic cannot advise against recognised customs and institutions: 'In practice, scepticism means a sympathetic stance toward the traditional and mistrust toward all utopias'.[60] Sensualistically oriented and rooted in experience, philosophical scepticism identifies '[i]mmediate perception and reflection, natural need, laws and tradition, honed

56 See Maria Antonietta Falchi Pellegrini, 'Il Montaigne di Horkheimer', *Il pensiero politico*, 22.2 (1989), 248–63 (pp. 249–50).

57 Horkheimer, 'Montaigne und die Funktion der Skepsis', pp. 240–1; 'Montaigne and the Function of Skepticism', p. 269 (translation amended).

58 Horkheimer, 'Montaigne und die Funktion der Skepsis', p. 241; 'Montaigne and the Function of Skepticism', p. 269.

59 Horkheimer, 'Montaigne und die Funktion der Skepsis', pp. 241–2; 'Montaigne and the Function of Skepticism', p. 269.

60 Horkheimer, 'Montaigne und die Funktion der Skepsis', p. 242; 'Montaigne and the Function of Skepticism', p. 270.

dexterity and customary knowledge' as guides for action. Ultimately, 'it is, in essence, conservative'.[61]

In Montaigne, the conservative element is even more pronounced than in the ancient sceptics insofar as he witnesses the rise of an absolutism capable of guaranteeing bourgeois property, thus also securing its interests. In this view, the nation-state looms as the institution that will establish a new order and protect the rising bourgeoisie.[62] In this way, ancient ataraxia is remodelled in Montaigne as 'the contended ordering of the inner spirit, in which one recovers from any inequity'; and *epoché* takes the form of 'the retreat into private interiority, in which one can replenish oneself, freed from the compulsions of occupational duties'.[63] This attitude reflects a phenomenon that, according to Horkheimer, characterises the beginning of the bourgeois era, namely, the demarcation, both in the individual and in the social totality, between the cultural and economic spheres: while the principle of duty and the laws of economic competition reign in labour and the economy, an eternal harmony prevails in the cultural universe.[64] This observation, however, applies to Montaigne only up to a point, as his retreat to his chateau and his practice of meditative solitude did not bring him immediate serenity and balance, but rather confrontation with inner monsters and ghosts.[65] This aspect of Montaigne's idleness could therefore be interpreted as the intrusion of the outside into the inside, as evidence of the difficulty of achieving a total separation between

61 Horkheimer, 'Montaigne und die Funktion der Skepsis', p. 243; 'Montaigne and the Function of Skepticism', pp. 270–1.

62 Horkheimer, 'Montaigne und die Funktion der Skepsis', p. 243–4; 'Montaigne and the Function of Skepticism', p. 271.

63 Horkheimer, 'Montaigne und die Funktion der Skepsis', p. 244; 'Montaigne and the Function of Skepticism', p. 271.

64 Horkheimer, 'Montaigne und die Funktion der Skepsis', p. 244; 'Montaigne and the Function of Skepticism', p. 271.

65 See for example the chapter I, 8: Michel de Montaigne, *Les Essais*, ed. by Pierre Villey and Verdun-Louis Saulnier (Paris: PUF, 2004), 'De l'oisiveté', pp. 32–3; *The Complete Essays of Michel de Montaigne*, trans. by Donald M. Frame (Stanford: Stanford University Press, 1958 [1948]), 'Of idleness', pp. 20–1.

the sphere of culture and reflection and the sphere of commitments (work and home) that Montaigne also seeks to establish.[66]

According to Horkheimer's reading, the entire life of the educated bourgeois at the dawn of the modern age is split into two spheres: profession on one hand, and distraction and edification on the other. While the former includes duties to one's family and responsibilities as a citizen, the latter is more precisely characterised as the realm of 'passing of time'.[67] This characterisation adds precision to the concept of modern scepticism: ataraxia is not limited to mere psychic interiority, but extends to general leisure and an experience of the world that is not torn apart by the competition and tensions that prevail in the economic sphere. Horkheimer, then, comes to this conclusion: 'Worldly enjoyment and retreat into interiority are identical for Montaigne. Those who sit in their libraries or who take a pleasant trip rest contentedly by themselves. The social stratum in France to which he belonged had the means to create a pleasurable private life'.[68] The condition of the masses of the underpriviledged was somewhat different, driven, as they were, by the crisis of the feudal order to seek new and burdensome work in the manufactories and to subject themselves to a new discipline of labour: individuals belonging to the lowest social strata withdrew into an interiority that was by no means a source of rest and pleasure, but in which they found,

66 See a passage from Chapter I, 39, evoked by Horkheimer himself: 'We should have a wife, children, goods, and above all health, if we can; but we must not bind ourselves to them so strongly that our happiness depends on them. We must reserve a back shop all our own, entirely free, in which to establish our real liberty and our principal retreat and solitude. Here our ordinary conversation must be between us and ourselves, so private that no outside association or communication can find a place; here we must talk and laugh as if without wife, without children, without possessions, without retinue and servants, so that, when the time comes to lose them, it will be nothing new to us to do without them'. Montaigne, *Les Essais*, 'De la solitude', p. 241; *Essays*, 'Of solitude', p. 177 (translation amended).

67 Horkheimer, 'Montaigne und die Funktion der Skepsis', p. 245; 'Montaigne and the Function of Skepticism', p. 272.

68 Horkheimer, 'Montaigne und die Funktion der Skepsis', p. 246; 'Montaigne and the Function of Skepticism', p. 273.

with the complicity of Protestantism, 'their own strict conscience, which accuses them of sin, explores their mistakes and oversights, and drives them to further work'.[69] However, the possibility of enjoying the tranquillity of resting in oneself, of a prosperity and culture that are still 'the fruits of a declining feudal order' contrasts with the new bourgeois order being established, in which 'culture exists only on the basis of the capitalist form of labour'.[70] This does not negate the tensions that arise in the relationship with the world, nor does it invalidate the awareness of the suffering that pollutes human relations, or the development of a feeling of empathy toward the poor and dispossessed. In Montaigne, the thinker who became a model for the educated man during the bourgeois era, Horkheimer also perceives expressions of 'solidarity with the oppressed'[71] as well as the courage to denounce the barbaric justice practised by Christians in the chapter 'Of cruelty' and European barbarism in 'Of cannibals'.[72]

Conversely, ataraxia and inner withdrawal find their theoretical counterpart in subjectivism. In a changing and deceptive world, whose very existence is not certain but must be demonstrated through complex reasoning, the self is alone; yet, in this solitude, it finds its strength, since 'the isolated self [*das einsame Ich*]' reveals itself as 'the only comprehensible reality [*die einzige verständliche Realität*]'.[73] In Horkheimer's synthesis of the tendencies of modern philosophy that emerge in contrast to Aristotelian scholasticism, the Cartesian discursive procedure is foreshadowed; yet Montaigne is tucked away therein: his 'style of thought [...] fits in with modern

69 Horkheimer, 'Montaigne und die Funktion der Skepsis', pp. 246–7; 'Montaigne and the Function of Skepticism', p. 273.
70 Horkheimer, 'Montaigne und die Funktion der Skepsis', p. 247; 'Montaigne and the Function of Skepticism', p. 273.
71 Horkheimer, 'Montaigne und die Funktion der Skepsis', p. 257; 'Montaigne and the Function of Skepticism', p. 281.
72 Horkheimer, 'Montaigne und die Funktion der Skepsis', p. 257, n. 60; 'Montaigne and the Function of Skepticism', p. 412, n. 62.
73 Horkheimer, 'Montaigne und die Funktion der Skepsis', p. 254; 'Montaigne and the Function of Skepticism', p. 279.

philosophy's concept of knowledge'.[74] The statement is noteworthy and partly explains why Horkheimer goes back to Montaigne during the 1930s to explore his personality and thought: firstly, the *Essays* constitute a foundational text of modern thought and, secondly, Montaigne's scepticism, by taking the individual as its positive content and showing that the *ego*, despite its misery and inconstancy, remains the only anchor to cling to both in theory and in practice, contributes decisively to establishing and nourishing the philosophical reflection of modernity.[75]

On a strictly practical level, the *ego*, being the only principle on which to rely, implies that the happiness of the individual depends on oneself. On the other hand, true wisdom consists in pursuing happiness to the extent that we can afford it, given our physical constitution, our temperament and external circumstances. 'The maxim is to act naturally. Violence against oneself and others, against people and animals, is foolhardiness. There is no logical proof against tyranny and cruelty'.[76] These lines effectively summarise Montaigne's ethical-practical orientation and, although the reference is fleeting, they evoke the presence in Montaigne of an ethics of nature and animals, a sphere of problems in which Horkheimer – as has already been emphasised – manifests a keen interest.[77]

The 1938 essay on Montaigne is also interesting insofar as Horkheimer highlights the points of contact between modern scepticism and Protestantism, while also being careful to point out some crucial divergences between the two movements.

Protestantism and Montaigne converge first and foremost in their criticism of knowledge and reason. Luther and Calvin condemn the human claim to rely on theoretical thought: instead of believing that human beings can attain a more exact and complete knowledge of

---

74 Horkheimer, 'Montaigne und die Funktion der Skepsis', p. 255; 'Montaigne and the Function of Skepticism', p. 279.

75 Horkheimer, 'Montaigne und die Funktion der Skepsis', p. 255; 'Montaigne and the Function of Skepticism', pp. 279–80.

76 Horkheimer, 'Montaigne und die Funktion der Skepsis', pp. 255–6; 'Montaigne and the Function of Skepticism', p. 280.

77 See Orietta Ombrosi, 'Le *miserere* des bêtes. Max Horkheimer et Theodor W. Adorno face à l'animalité', *Lignes*, 28.1 (2009), 146–66.

the truth and identify the criterion of action starting with themselves, they must submit to the word of God and authority. It can be asserted that Montaigne acts as a counterpoint to this position when – in the 'Apology for Raymond Sebond' – he questions the role and value of human reason and knowledge, not only in theology, but also in science.[78] 'The plague of man is the opinion of knowledge'[79] is one of the passages quoted by Horkheimer as evidence of certain strong tones in which Montaigne demystifies the belief in human knowledge and warns against the weakness and danger of the individual's rationalistic claims.[80]

Horkheimer quotes, in particular, two passages from the 'Apology for Raymond Sebond', the second of which is presented as the outcome of reflection that accepts the sceptical point of view (united here as a single quotation):

> [C]People who judge and check their judges never submit to them as they ought. How much more docile and easily led, both by the laws of religion and by political laws, are the simple and incurious minds, than those minds that survey divine and human causes like pedagogues!
>
> [A]There is nothing in man's invention that has so much verisimilitude and usefulness. It presents man naked and empty, acknowledging his natural weakness, fit to receive from above some outside power; stripped of human knowledge, and all the more apt to lodge divine knowledge in himself, [B]annihilating his judgment to make more room for faith; [C]neither disbelieving [A]nor setting up any doctrine [B]against the common observances; humble, obedient, teachable, zealous; a sworn enemy of heresy, [A]and consequently free from the vain and irreligious opinions introduced by the false sects. [B]He is a blank tablet prepared to take from the finger of God such forms as he shall be pleased to engrave on it.[81]

78 Horkheimer, 'Montaigne und die Funktion der Skepsis', p. 247; 'Montaigne and the Function of Skepticism', pp. 273–4. See, for example, Montaigne, *Les Essais*, II, 12, 'Apologie de Raymond Sebond', p. 450; *Essays*, 'Apology for Raymond Sebond', p. 329.

79 'La peste de l'homme, c'est l'opinion de sçavoir'. Montaigne, *Les Essais*, II, 12, p. 488; *Essays*, p. 360.

80 Horkheimer, 'Montaigne und die Funktion der Skepsis', p. 248; 'Montaigne and the Function of Skepticism', p. 274.

81 Montaigne, *Les Essais*, II, 12, p. 506; *Essays*, p. 375.

In the second part of the quote, Montaigne argues that human beings find themselves in the paradoxical situation of being deprived of human science yet able to accommodate within themselves divine science, to the extent that they recognise their own weakness. It is, therefore, a verisimilar and useful idea that presents the human being as naked, empty and fragile: it makes individuals humble, obedient, more likely to follow general customs and habits, inclined to be guided, and immune to the contagion of heresies. Horkheimer notes that Montaigne intends to criticise the sects here, and that his attacks are certainly aimed at the Huguenots, yet they are not intended to strike at their faith but at their claim to be a French party threatening the quiet life of state and society. In fact, according to Horkheimer, Montaigne's attitude is consistent with that of 'a triumphant Protestantism' and, more precisely, with the dominant tendency in the republic of Geneva shaped by Calvinism, where humility was more esteemed than vanity and faith was preferred to science.[82] According to Luther, on the other hand, human beings must resign themselves to absolute passivity; moreover, compared to the latter's assertion of the presumption of human reason, Montaigne's sceptical *epoché* proves weak, and Calvin himself appears to be a Catholic rather than a reformer.[83] However, according to Horkheimer, both Montaigne and the Protestants do not reject thought and science as such, but only to the extent that they come into conflict with the established order, that is, with institutions, norms, and customs that have a legitimate existence, that have taken root in a certain context and are thereby approved.[84]

Nevertheless, Montaigne radically departs from Protestantism on one decisive issue. According to Horkheimer, as opposed to the Protestant Reformers' attitude, Montaigne highlights moderation as the highest virtue, rather than absolute self-abnegation. He views the antagonistic parties – in this case Catholics and Reformers –

82 Horkheimer, 'Montaigne und die Funktion der Skepsis', p. 249; 'Montaigne and the Function of Skepticism', p. 275.

83 Horkheimer, 'Montaigne und die Funktion der Skepsis', p. 249; 'Montaigne and the Function of Skepticism', p. 275.

84 Horkheimer, 'Montaigne und die Funktion der Skepsis', p. 250; 'Montaigne and the Function of Skepticism', p. 276.

from the standpoint of an enlightened diplomat. In short, for him, no one is in the right; indeed, there is no ultimate, universal standard of right and wrong, but only states of order or disorder.[85] The question of moderation is central here: in Chapter I, 30 (a text later quoted by Horkheimer) Montaigne writes: that one can love virtue immoderately and devote oneself too zealously to a just action; that the awareness of the danger inherent in excesses (even in the good) leads him to prefer 'temperate and moderate natures'.[86] In this sense of measure, as in the *epoché* practiced by Montaigne, linked to the reservation of judgement and inner calmness, Horkheimer perceives 'a historical reason', a desire to keep oneself 'free from the historical upheaval', and 'a progressive attitude' for the sixteenth century: 'such moderation was identical with the preservation of one's own person, with going the objectively correct way of tolerance in the national state. It amounted to independence from the illusions of the religious parties'.[87]

Thus, with a dialectical movement, Horkheimer shows that while, at first, and on certain points in particular (such as the critique of the human claim to knowledge), Montaigne's position converges with that of Protestantism, a considerable distance later opens up between the Reformers and Montaigne. The latter's moderation – which leads him, as has already been emphasised, to express solidarity with the oppressed – stands in stark contrast to certain pinnacles of inhumanity and the dark intransigence of the former.[88]

Horkheimer, like Erich Fromm a few years later,[89] emphasises the link between the economic transformations that mark early modernity, existential concerns, and the reform movement. In this regard, he writes:

---

85 Horkheimer, 'Montaigne und die Funktion der Skepsis', p. 250; 'Montaigne and the Function of Skepticism', p. 276.

86 See Montaigne, *Les Essais*, I, 30, 'De la modération', p. 197; *Essays*, 'Of moderation', p. 146.

87 Horkheimer, 'Montaigne und die Funktion der Skepsis', p. 269; 'Montaigne and the Function of Skepticism', p. 291.

88 Horkheimer, 'Montaigne und die Funktion der Skepsis', p. 257; 'Montaigne and the Function of Skepticism', pp. 281–2.

89 Fromm, *Escape from Freedom*, pp. 80–1.

> Mediated by their fanaticism, there emerges the bourgeois mass individual, who grows out of the childish condition of the medieval individual through the inversion of material desires, the subjugation of sensual impulses to a relentlessly driving ego, and the psychic incorporation of economic and political pressure as duty. Such individuals adapt to the nascent bourgeois order but with rancor and strong faith, with jealousy and guilty feelings, with sexual envy and misanthropy.[90]

In the doctrines of the Reformers 'God and world, freedom and servitude, natural instinct and conscience, divine and earthly commandment remain [...] unresolved contradictions, obscuring the real contradiction of the individual who seeks to develop itself with the relations of an emergent capitalism'.[91] After all, as Erich Fromm points out, analysing the religious ideas of early modernity from the perspective of their psychological significance, Protestantism gives expression, on a theological level, to the paradoxical condition of modern man, freed from the bonds of medieval orders and now entrusted only to himself. On the one hand, modern man perceives himself to be independent of external authorities; on the other, he feels a growing isolation and the consequent feeling of irrelevance and powerlessness.[92] Fromm, not surprisingly, emphasises that Lutheran doctrine has two faces: the critique of authority and the individual's condition of isolation.[93] By emphasising the individual's radical weakness and inherent wickedness, Lutheranism states that, only if they humble themselves and destroy their individual will, can they hope to receive God's grace.[94] The Protestants preach the deferral of the individual's material impulses; they establish a sort of introversion of the aspirations of the masses, diverting 'the demands of the dominated away from the rulers and toward their own inner nature'. 'Economic compulsion' is 'mystified', transfigured, brought

90 Horkheimer, 'Montaigne und die Funktion der Skepsis', pp. 257–8; 'Montaigne and the Function of Skepticism', p. 282 (translation amended).
91 Horkheimer, 'Montaigne und die Funktion der Skepsis', p. 259; 'Montaigne and the Function of Skepticism', p. 283 (translation amended).
92 Fromm, *Escape from Freedom*, p. 38.
93 Fromm, *Escape from Freedom*, p. 74.
94 Fromm, *Escape from Freedom*, p. 75.

back to the inscrutable divine will:[95] Horkheimer therefore sees, in the consolidation of the Reformation, the affirmation of an economic theology legitimising social inequalities and silencing the demand of the oppressed for an existence worthy of man. On this point, too, Montaigne's position does not align with the Protestant orientation. He detests oppression in both the public and private spheres but admits that he has no intention of taking on the task of abolishing injustice.[96]

It is no coincidence that Horkheimer quotes the following passage from the last chapter of the *Essays* at a central point in his article:

> I am so sick for freedom, that if anyone should forbid me access to some corner of the Indies, I should live distinctly less comfortably. And as long as I find earth or air open elsewhere, I shall not lurk in any place where I have to hide. Lord, how ill could I endure the condition in which I see so many people, nailed down to one section of this kingdom, deprived of the right to enter the principal towns and the courts and to use the public roads, for having quarreled with our laws! If those that I serve threatened even the tip of my finger, I should instantly go and find others, wherever it might be.[97]

'These words – Horkheimer comments – express not only reactionary indifference [*reaktionäre Gleichgültigkeit*], but also a revolutionary humanism as well [*ein revolutionärer Humanismus ausgedrückt*]. Today's sceptics don't leave the country as long as the bureaucracy lets them be. In any case, it makes no sense to leave, for "earth and air" are no longer open anywhere, and that stretch of land in the Indies is ruled by the same laws that one seeks to flee'.[98] Further on, Horkheimer states that the humanity

---

95 Horkheimer, 'Montaigne und die Funktion der Skepsis', p. 259; 'Montaigne and the Function of Skepticism', p. 283.

96 Horkheimer, 'Montaigne und die Funktion der Skepsis', p. 261; 'Montaigne and the Function of Skepticism', p. 284.

97 Montaigne, *Les Essais*, III, 13, 'De l'expérience', p. 1072; *Essays*, 'Of experience', pp. 820–1. The passage is quoted in Horkheimer, 'Montaigne und die Funktion der Skepsis', pp. 270–1; 'Montaigne and the Function of Skepticism', p. 292.

98 Horkheimer, 'Montaigne und die Funktion der Skepsis', p. 271; 'Montaigne and the Function of Skepticism', p. 292 (translation amended).

that expresses itself in cosmopolitanism and diplomacy, typical of the ancients, Montaigne, and even Hume, now dispenses with its peaceful vocation; 'at present, that attitude reveals only the desire to participate a little in a decadent power'.[99] Horkheimer grasps in Montaigne a dialectical polarity between a reactionary tendency and a cult of subjectivity on the one hand and a revolutionary charge, an open and cosmopolitan humanism on the other.[100]

> Montaigne's *Epoché* – as Horkheimer writes elsewhere in his essay – is not without solidarity with humanity. To practice it entails promoting the happiness with which he was preoccupied for the individual ego, not just in particular, but in general. At present, however, only the futility of the principle is evident. The peace that the liberal sceptic today concludes with the authoritarian order expresses not humane praxis but its renunciation'.[101]

By means of a continuous 'movement' from the past to the present (and vice versa) and a close comparison between Montaigne's scepticism and that of the early twentieth century, Horkheimer's article thus unveils the different sides of the scepticism of the early modern age: while at first focusing on its individualistic charge and its conservatism, at a later stage it highlights its humanistic,

99 Horkheimer, 'Montaigne und die Funktion der Skepsis', p. 272; 'Montaigne and the Function of Skepticism', p. 293 (translation amended).

100 On this topic, see Ursula Reitemeyer, 'Die Funktion der Skepsis: Montaigne zwischen Tradition und Modern', in Ursula Reitemeyer, *Umbruch in Permanenz. Eine Theorie der Moderne zwischen Junghegelianismus und Frankfurter Schule* (Münster: Waxmann, 2007), pp. 28–41 (p. 41). On Horkheimer and Montaigne see also Vicente Raga, 'Tentativas sobre Montaigne: Horkheimer y la función del escepticismo', *Estudios de Filosofía*, 55 (2017), 82–102; Susanne Zepp, 'Max Horkheimer: Montaigne und die Funktion der Skepsis', in *Textgelehrte. Literaturwissenschaft und literarisches Wissen im Umkreis der Kritischen Theorie*, ed. by Nicolas Berg and Dieter Burdorf (Göttingen: Vandenhoeck & Ruprecht, 2014), pp. 81–90; Raffaele Carbone, 'Umanesimo, scetticismo, società borghese: Horkheimer lettore di Montaigne', *Atti dell'Accademia delle Scienze morali e politiche*, 128 (2018), 7–40.

101 Horkheimer, 'Montaigne und die Funktion der Skepsis', p. 270; 'Montaigne and the Function of Skepticism', pp. 291–2.

revolutionary and demystifying side of the feudal order, i.e. the illusions prevailing in the transition phase to modernity. In continuity with 'Egoism and Freedom Movement', the essay on Montaigne explores the nuances and chiaroscuros of the dawn of the modern age, but also the transformations that occurred throughout modernity within the very bosom of bourgeois society. The alliance between the bourgeoisie and absolutism at the beginning of the modern era, which is embodied precisely in Montaigne's attitude, is the result – as Horkheimer writes – of the process of 'emancipation of the bourgeoisie from a bankrupt feudalism' and cannot be equated with the alliance between the bourgeoisie and fascism in the 1920s and 1930s. This alliance was born of fear of the proletariat and causes a deformation of sceptical tolerance towards freedom of conscience, which can subsequently turn into 'conformism with the regime of the secret police'.[102] In short, from a humanistic mental attitude, scepticism transforms into pure conformism, and this process has its roots in the economic configuration of the time.[103]

Horkheimer then draws attention to an error of perspective that, already present in humanistic scepticism, becomes progressively clearer in the developments of modern scepticism, particularly since Hume. The independence of the self, presupposed by scepticism, rests on the individual freedom that each economic subject should enjoy in a market economy. In reality, since labour and the means of production are socially divided, that is, distributed among different classes, each of which is rooted in a specific relationship to the means of production, '[t]he equality of free individuals, which renews itself through exchange, with each person's labour as the basis of their possessions and power, in short, the principle of the bourgeoisie – upon which its ideology, justice, and morality rest – [...] reveals itself as a mere facade that masks the true relations'.[104] Individuals are truly free if they are not deprived of their freedom by economic

102 Horkheimer, 'Montaigne und die Funktion der Skepsis', p. 270; 'Montaigne and the Function of Skepticism', p. 292.

103 Horkheimer, 'Montaigne und die Funktion der Skepsis', p. 273; 'Montaigne and the Function of Skepticism', p. 294.

104 Horkheimer, 'Montaigne und die Funktion der Skepsis', pp. 273–4; 'Montaigne and the Function of Skepticism', p. 295 (translation amended).

processes and their political consequences. Horkheimer thus perceives a certain ambiguity – which, in reality, reveals an internal contradiction within the bourgeois concept of freedom[105]– in the 'modern sceptical attitude' insofar as it rejects revolutionary action and the 'critique of the totality': its main characteristics, namely liberality, subjectivism and relativism, conceal 'a harsh, misanthropic quality; it is not as just and open as it sometimes appears'.[106] The dark side of scepticism reveals itself when the ideals of freedom and equality produced by bourgeois culture are confronted with the harsh reality of economic liberalism. If Horkheimer's text is interpreted correctly, this dark side appears in a diachronic dimension, reflecting the developments of the modern world: at the dawn of modernity, scepticism presents humanistic traits and questions both the feudal socio-economic order and the medieval worldview; at the end of the sixteenth century, it begins to turn into pure conformism until it becomes cynically and enthusiastically fascistic in its orientation.[107] This type of scepticism fails to recognise that, within the history of bourgeois society, behind its ideals lies the cruelty of economic laws. Moreover, by defending the idea that theory is relative and divorced from practice, it undermines and thwarts any project aimed at transforming the existing social order.[108]

Horkheimer opposes this attitude with the active humanism of critical theory, which is originally rooted in the link between humanism and sixteenth-century scepticism.

> What set the Reformation and scepticism against each other in the sixteenth century – fanatical spontaneity on the one hand and humanism on the other – has been released from these forms and transformed into a theory and praxis which, as active humanism [*als aktiver Humanismus*], overcomes and retains the contradiction [*den Gegensatz überwindet*

105 Horkheimer, 'Montaigne und die Funktion der Skepsis', p. 274; 'Montaigne and the Function of Skepticism', p. 295.

106 Horkheimer, 'Montaigne und die Funktion der Skepsis', p. 274; 'Montaigne and the Function of Skepticism', p. 295 (translation amended).

107 Horkheimer, 'Montaigne und die Funktion der Skepsis', p. 278; 'Montaigne and the Function of Skepticism', p. 299.

108 Horkheimer, 'Montaigne und die Funktion der Skepsis', p. 275; 'Montaigne and the Function of Skepticism', p. 296.

> *und bewahrt*]. It is critical theory, and the historical effort to which this belongs. It is to be found concretely among those in authoritarian states (and those striving to become such) who build the cells of a new world. To them, even after the defeat, thought has not become an internal matter that remains confined and adapts to a contradictory reality. They do not wash their hands in innocence. It is possible that everything will go to rack and ruin, but even the most sober analysis demonstrates that a rational society is possible. Humanism consists in committing oneself to its creation [*Humanismus besteht darin, sich für sie einzusetzen*].[109]

Opposing the closure of people to one another, 'active humanism [*der aktive Humanismus*]' as it has developed throughout history now plays the role once assigned to sceptical philosophers and reformers.[110] Critical theory incorporates and thereby transcends the opposition between that spontaneous, active dimension aimed at influencing real practices, which is characteristic of the Reformers (thus neutralising the charge of fanaticism brought against them) and the humanistic component of scepticism that encapsulates a theoretical–reflexive dimension and an authentic openness to the various forms of the human, including a sense of solidarity with the oppressed.

In these pages, Horkheimer presents critical theory as a new form of humanism that transcends the limits of early modern humanism and

109 Horkheimer, 'Montaigne und die Funktion der Skepsis', p. 289; 'Montaigne and the Function of Skepticism', p. 307 (translation amended).

110 Horkheimer, 'Montaigne und die Funktion der Skepsis', p. 290; 'Montaigne and the Function of Skepticism', p. 308. Scholars have noted that Horkheimer rarely uses the expression 'active humanism'. Nevertheless, it is a fundamental idea. See Alfred Schmidt, Gian Enrico Rusconi, *La Scuola di Francoforte: Origini e significato attuale* (Bari: De Donato, 1972), pp. 40–1. As Falchi Pellegrini points out, it alludes to a humanism that does not seek to protect the self, nor does it constitute an abstract idealisation of man, but aims to be a rational commitment to making society more human. See Falchi Pellegrini, 'Il Montaigne di Horkheimer', p. 261. According to Oliver Kozlarek, 'Horkheimer's proposal of a practical, "active" actualization and appropriation of humanist culture, activated through critical social research, is to be understood as a response to the ideological abuse of humanistic ideas'. Oliver Kozlarek, *The Critical Humanism of the Frankfurt School as Social Critique* (Lanham, MD: Lexington Books, 2024), p. 73.

distinguishes itself from the new humanism theorised by Siegfried Marck[111] which, in opposition to national socialism, presents the traits of a non-Marxist, non-materialistic socialism that would not disdain the connotation of 'idealistic' if the concept of idealism had not become empty and abstract.[112] In the aforementioned article 'The Philosophy of Absolute Concentration', Horkheimer responds to Marck's neo-humanism – which incorporates the German cultural tradition (from Goethe to Thomas Mann) as well as various cultural tendencies of his time (such as Gestalt philosophy) and which considers it possible, by mobilising this heritage, to restore humanity to itself without excluding a 'friendship with being [*die Freundschaft mit dem Sein*]' – by stating that 'until the world is transformed from its very foundations, there can be no friendship with any being'.[113]

The humanism of early modernity, on which some pages of the essay on Montaigne focus, is characterised as rationalistic–bourgeois. While carrying an erosive charge against the Scholastic worldview and the feudal order, it tends to retreat into interiority and ends up suggesting, in the involutional development of bourgeois society, that men adapt themselves, as best they can, to reality, even when said reality is at odds with their aspirations.[114] According to

111 At the end of 1937, together with Heinrich Mann, Georg Bernhard, Emil Julius Gumbel and Fritz Lieb, who all knew each other from the Popular Front movement, Siegfried Marck founded the 'League of Free Socialists' in Dijon. The 'Bund freiheitlicher Sozialisten' aimed to reconcile the values of political liberalism (especially the idea of freedom) with the goals of socialism. Its founders believed that fascism succeeded because it separated the partisans of freedom and the workers advocating socialism by defeating them individually. Considering that the lack of a united bourgeois organisation was also a key reason for the failure of the Popular Front, their goal was to strengthen the bourgeois component in the fight against national socialism without renouncing the principles of 1789. See, in this regard, Markus Schultz's doctoral thesis, 'Siegfried Marck: Politische Biographie eines jüdisch-intellektuellen Sozialdemokraten' (PhD dissertation, University of Göttingen, 2016), pp. 175–7.

112 These are Marck's own ideas evoked by Horkheimer in the article 'Die Philosophie der absoluten Konzentration', p. 299.

113 Horkheimer, 'Die Philosophie der absoluten Konzentration', p. 300.

114 For more on this point, see Antonio Ponsetto, *Max Horkheimer. Dal mito della distruzione alla distruzione del mito* (Bologna: il Mulino, 1981), p.

Horkheimer, the updated humanism of critical theory does not claim to immunise the pragmatic bourgeois against barbarism by resorting to tradition, ancient Greece or even, on occasion, Thomism: such a humanism bears no relation to the real world.[115] In other words, active humanism does not entrench itself within the confines of philological and hermeneutic engagement with the Western literary and philosophical tradition, nor does it erect a barrier between the self and the world, much less take refuge in an idealised image of the individual. It is commensurate with the ability to untie the problematic knots of one's time. Indeed, Horkheimer continues, 'not just any old ideas but the true ones, in their historically adequate form [*in ihrer historisch adäquaten Gestalt*], distinguish cultural development [*Bildung*] from mere knowledge [*vom bloßen Wissen*]'.[116] Humanism that truly shapes the individual is not reduced to a mere philological–speculative activity but consists of intellectual work attuned to the spirit of the age, in the elaboration of *wahren Ideen* that enable one to understand an epoch and to create the conditions for dealing with its critical aspects. Critical theory embodies this ideal of humanism. It takes up the legacy of early modern humanism and enriches it with the resources of materialist dialectics. In this way, it 'rejects the unity of thought and history', revealing the absence of harmony between thought and being, as well as the contradiction that exists 'between human beings and nature' and that 'between human beings with their needs and capacities and society'.[117] As can be clearly seen, here Horkheimer does not address the nexus between the project of transforming society and the idea of redefining the relationship between human beings and nature in the light of active humanism, but suggests that there is a dialectical and historical relationship between nature and human societies.

173.

115 Horkheimer, 'Montaigne und die Funktion der Skepsis', p. 290; 'Montaigne and the Function of Skepticism', p. 308.

116 Horkheimer, 'Montaigne und die Funktion der Skepsis', p. 290; 'Montaigne and the Function of Skepticism', p. 308.

117 Horkheimer, 'Montaigne und die Funktion der Skepsis', p. 291; 'Montaigne and the Function of Skepticism', p. 309.

Returning to the topic of humanism, it is important to note that dedicating oneself to culture – as Horkheimer states in a university speech delivered in 1952 – does not mean practising the cult of the self-sufficient Ego but recognising 'the connection it has with the whole'.[118]

> There is no humanism without a clear position toward the historical problems of the epoch; it cannot exist as a mere profession of faith in itself. The humanism of the past consisted in the critique of the hierarchical feudal order, which had become a fetter on the development of humanity. The humanism of the present lies in the critique of the forms of life [*in der Kritik der Lebensformen*] under which humanity now perishes, and in the effort to transform them in a rational manner [*in vernünftigem Sinn*].[119]

In this passage, confirming the coherence of Horkheimer's 1930s writings and the thread running through them, the fundamental concepts highlighted in 'Traditional Theory and Critical Theory' resonate: the need 'to grasp in thought the unity of theory and practice',[120] the critique of present society, and the related project of transforming that society into one that is 'rationally organised'.[121]

---

118 Horkheimer, 'Begriff der Bildung', in Horkheimer, *GS* 8: *Vorträge und Aufzeichnungen 1949-1973*, ed. G. Schmid Noerr (Frankfurt am Main: Fischer, 1985), pp. 409–19, here pp. 415, 416, respectively.

119 Horkheimer, 'Montaigne und die Funktion der Skepsis', p. 290; 'Montaigne and the Function of Skepticism', p. 308 (translation amended). It should be noted that Horkheimer does not view historical Humanism as an exclusively literary–cultural phenomenon: the polemic against the decadence of the Latin language in the Middle Ages and scholastic obscurity, in his view, translates into a critique of the feudal system and its hierarchical structure.

120 Horkheimer, 'Traditionelle und kritische Theorie', p. 205; 'Traditional and Critical Theory', p. 231.

121 Horkheimer, 'Traditionelle und kritische Theorie', p. 207; 'Traditional and Critical Theory', p. 233 (translation amended). See also 'Traditionelle und kritische Theorie', pp. 214–15; 'Traditional and Critical Theory', p. 241. The idea of transforming society in order to improve human existence, in other words, to free the majority of individuals from subjugation and suffering, is found in several of Horkheimer's texts, for example, in 'Materialism and Morality' – even after the writings of the

From the perspective of critical theory, humanism cannot crystallise into a pure *Bekenntnis*, a profession of faith in one's own idea, perpetually renewed in theory yet detached from the historical context of its time. It is no coincidence that, in order to clarify what humanism cannot be, Horkheimer uses a term, '*Bekenntnis*', which, among other things, carries a religious semantic connotation: humanistic practice cannot retreat into itself, confined to pure philological-hermeneutic work on the texts of the past – just as a certain religiosity is expressed in the believer's exclusive relationship with God through the practice of confession –, but must become aware of the historical problems of its own time and expose the contradictions and criticalities of its forms of life.

As is well known, a significant landmark in the project of critical theory came in 1941 with 'The End of Reason', following the calamitous change in the historical and theoretical horizon, marked by the rise of a radically planned state capitalism in Germany and the Soviet Union, which led Horkheimer to reformulate his idea and the call for a rational society. Nevertheless, it should also be noted that the idea of humanism, outlined in the essay on Montaigne, returns in *Späne*, the annotations drafted by Pollock. Here, in a 1952 aphorism, entitled 'Negativer Humanismus', Horkheimer states – as does Adorno in his 1963 course on *Problems of moral philosophy*[122] – that it is possible to determine, not the nature of man (understood in a qualitative sense), but what is inhuman.[123] Nevertheless, in

---

1930s. See a reflection from the years 1957–58, entitled 'Humanity': 'Our practical philosophy is humanity [*die Humanität*]. That men do not suffer misery, that created beings can develop, is the purpose of action in general'. Horkheimer, *Notizen 1949-1969*, p. 259; *Decline. Notes 1950-1969*, p. 152.

122 Theodor W. Adorno, *Probleme der Moralphilosophie*, ed. by Thomas Schröder (Frankfurt am Main: Suhrkamp, 2010), p. 261; *Problems of Moral Philosophy*, trans. by Rodney Livingstone (Cambridge: Polity Press, 2000), p. 175.

123 'Die Wahrheit läßt sich nicht bestimmen (man versuche eine Definition), wohl aber die Lüge. Bestimmen läßt sich nicht das Wesen des Menschen, wohl aber was unmenschlich ist, nicht das Gute, wohl aber was nicht gut ist. Die Funktion des Denkens ist die "bestimmte Negation". Das Negative und der Prozeß der "bestimmten Negation" sind die einzige Annäherung

the subsequent annotation, also from 1952, on 'Humanismus nach Montaigne und die Funktion der Skepsis', Horkheimer reiterates his conviction that active humanism can, in the twentieth century, take on the function that sceptical philosophers and reformers once did.[124] He then recalls the distinction, illustrated in the essay on Montaigne, between the humanism of the past and that of the present, highlighting their respective roles.[125] Finally, he reaffirms – in practically the same words as in the article on Montaigne – that authentic humanism cannot be reduced to a mere culture of interiority, but consists in taking a stand against the problems of one's own time.[126]

This problematic is also, and significantly, expressed in the inaugural speech for the re-opening of the Institute for Social Research in 1951. Here, Horkheimer emphasises the strength of the 'current humanism' that innervates social science[127] and points out that, in sociological work, an intention going beyond the existing society emerges. Moreover, he states that 'a critical attitude is inseparable

---

an die Wahrheit, die uns zugänglich ist. Dem, was sich als das Absolute aufspreizt, die Negation entgegensetzen, ist der Sinn des Glaubens'. Horkheimer, *Späne*, 'Negativer Humanismus', p. 200.

124 Horkheimer, *Späne*, 'Humanismus nach Montaigne und die Funktion der Skepsis', pp. 200–1.

125 'Der Humanismus der Vergangenheit bestand in der Kritik der feudalistischen Weltordnung mit ihrer Hierarchie, die zur Fesselung der Entfaltung des Menschen geworden war. Der Humanismus in der Gegenwart entfaltet die Kritik der Lebensformen, unter denen die Menschheit jetzt zu Grunde geht, in der Anstrengung, sie in einem vernünftigeren Sinn zu ändern'. Horkheimer, *Späne*, p. 201.

126 'Es gibt keinen Humanismus ohne klare Stellung zu den geschichtlichen Problemen der Epoche; als bloßes Bekenntnis zu sich selbst kann er nicht existieren'. Horkheimer, *Späne*, p. 201.

127 'Wir sehen in der Sozialwissenschaft ein Element jenes aktuellen Humanismus, mit dessen Entfaltung die Frage nach der Zukunft der Menschheit heute verbunden ist.' 'We see in social science an element of that current humanism with whose development the question of the future of humanity is now connected'. Institut für Sozialforschung an der Johann Wolfgang Goethe-Universität Frankfurt am Main, *Ein Bericht über die Feier seiner Wiedereröffnung, seine Geschichte und seine Arbeiten* (Frankfurt am Main: Stiftung Institut für Sozialforschung, 1952), p. 10.

from the profession of the social theorist'.[128] Humanism and critical attitude are thus intertwined as the Frankfurt Institute resumes its activities in Germany, in a discourse that outlines a clear vision of cultural education as a pathway through which the student learns how to deal with the 'tension with what exists'.[129] Social research contributes to the emancipation of individuals by making them politically aware subjects, capable of aspiring to a more just world founded on rational foundations. It aims to defuse that attitude, of a positivist nature, which is a prisoner of facts, of that which exists (a theme that would later return in *Notes 1949-1959*, linking up with the Marxist critique of the fetishisation of facts).[130]

In light of these considerations, critical theory is characterised, at least until the 1950s – and therefore before and after the 'turning point' marked by 'The End of Reason', and despite the remodulations by which it is distinguished over these decades – as a humanism of praxis: it aims to uproot the chains of superstition and misery that make individuals unhappy. On the other hand, in its theoretical and practical interest in emancipation, it draws on an essential tendency of the Western philosophical tradition (particularly German idealism, but also Greek philosophy 'in the golden age of Plato and Aristotle'),[131] i.e., on the process of rationalisation that runs through it, in line with the original Marxian position.[132] Here, however, Horkheimer consciously follows the Marxist tradition: the connection

128 *Ein Bericht über die Feier seiner Wiedereröffnung*, pp. 11–12.

129 *Ein Bericht über die Feier seiner Wiedereröffnung*, p. 12.

130 See Horkheimer, *Notizen 1949-1969*, 'Leere Vermittlung', 'An die Positivisten', pp. 295–6, 298–300; *Decline: Notes 1950-1969*, 'Empty Mediation', p. 164.

131 Horkheimer, 'Nachtrag' (1937), *GS* 4, pp. 217–25 (p. 219); 'Postscript', in Horkheimer, *Critical Theory: Selected Essays*, pp. 244–52 (pp. 245–6) (translation amended).

132 See Alfred Schmidt, *Die Kritische Theorie als Geschichtsphilosophie* (Munich: Hanser, 1976), pp. 83–4; Stefano Petrucciani, *Ragione e dominio. L'autocritica della razionalità occidentale in Adorno e Horkheimer* (Rome: Salerno editrice, 1984), pp. 28–9. Werner Post also emphasises this continuity, albeit in more muted tones. See Werner Post, *Kritische Theorie und metaphysischer Pessimismus. Zum Spätwerk Max Horkheimers* (Munich: Kösel Verlag, 1971), pp. 37– 8.

between classical German philosophy and the critical-emancipatory perspective of scientific socialism appears on several occasions in certain formulations by Marx and Engels and is significantly underlined by Karl Korsch in *Marxism and Philosophy*.[133] In this way, critical theory – at least in its first formulation in the 1930s and in the notes of the early 1950s – also ties in with the humanism of the early modern age, the limits and subsequent involution of which it overcomes – a humanism whose decline as a programme of human formation Horkheimer, in any case, notes in the 1950s: indeed, as Horkheimer affirms in a talk given in Regensburg in 1956 and published in 1957 with the title 'The Concept of Man', humanism gives way to the new technological sensibility and behaviour.[134] The project of a rationally organised society – in which the mythology of the necessity of economic data as immediate or natural facts is overcome and in which production is socialised and planned in a democratic manner – is thus the coherent outcome of an active humanism, a humanism that questions the existing power relations and is shaped as a programme for the emancipation of individuals from misery and subjugation.

Significantly, this conception of humanism is opposed to the dominance of subjective reason and the overflowing instrumental reason, 'its operational value, its role in the domination of men and nature', with the consequent reduction of thought 'to the level of industrial processes'; this ultimately leads to the reduction of reason itself to 'a fetish, a magic entity that is accepted rather

133 Karl Korsch, *Marxism and Philosophy*, trans. by Fred Halliday (London–New York: Verso, 2012), p. 30.

134 Max Horkheimer, 'Zum Begriff des Menschen' (1957), in Horkheimer, *GS* 7: *Vorträge und Aufzeichnungen 1949-1973*, ed. by Alfred Schmid and Gunzelin Schmid Noerr, 2nd edition (Frankfurt am Main: Fischer, 2014), pp. 55–80 (p. 64); 'The Concept of Man', in Horkheimer, *Critique of Instrumental Reason: Lectures and Essays Since the End of World War II*, trans. by Matthew J. O'Connell and others (New York: The Seabury Press, 1974), pp. 1–33 (p. 13) (translation amended). This involution manifests itself in the role of universities, which have been degraded to technical schools. For more on this point, see Oliver Kozlarek, 'From the Humanism of Critical Theory to Critical Humanism', *European Journal of Social Theory*, 24.2 (2020), 1–18 (p. 5).

than intellectually experienced'.[135] The critical element present in authentic humanism interpenetrates with philosophy's own vocation towards the current prevalence of subjective reason at the expense of objective reason – that type of reason that designates both 'a structure inherent in reality' and the 'very effort and ability to reflect such an objective order'.[136] According to Horkheimer, the task of philosophy is not to stand up for one of these two conceptions of reason, but 'to foster a mutual critique and thus, if possible, to prepare in the intellectual realm the reconciliation of the two in reality'.[137] Reason itself must exercise a form of self-critique that recognises the limits and one-sidedness of the two concepts of reason by analysing the moments through which the fracture between the two concepts was produced and developed. However, in our time, with the triumph of subjective reason, that is, the instrumental-formalist-subjectivist orientation, according to Horkheimer, it is necessary to recall and emphasise the worldview proper to objective reason.[138] It is then a matter of remembering that the objective conception of reason looked to a totality (and not only to human beings and their ends), to an objective structure that constitutes the touchstone by which the reasonableness of individual thoughts and actions can be assessed.

---

135 Horkheimer, *Zur Kritik der instrumentellen Vernunft*, *GS* 6, pp. 19–186 (pp. 42 and 44, respectively); *Eclipse of Reason* (New York: Oxford University Press, 1947), pp. 21 and 23, respectively.

136 Horkheimer, *Zur Kritik der instrumentellen Vernunft*, p. 34; *Eclipse of Reason*, p. 11.

137 Horkheimer, *Zur Kritik der instrumentellen Vernunft*, p. 175; *Eclipse of Reason*, p. 174.

138 Horkheimer, *Zur Kritik der instrumentellen Vernunft*, pp. 175–6; *Eclipse of Reason*, p. 174. Nevertheless, the possibility of a self-criticism of reason, in which Horkheimer grasps the true goal of philosophical thinking ('to essay a reconciliation [Versöhnung]'), is problematic (Horkheimer, *Zur Kritik der instrumentellen Vernunft*, *GS* 6, p. 165; *Eclipse of Reason*, p. 115). For example, if the true philosophy emphasises the difference between the ideal and the real, it produces 'at the same time a relapse into metaphysics'. Georg Lohmann, 'The Failure of Self-Realization: An Interpretation of Horkheimer's *Eclipse of Reason*', in *On Max Horkheimer: New Perspectives*, ed. by Seyla Benhabib, Wolfgang Bonß, and John McCole (Cambridge, MA–London: The MIT Press, 1993), pp. 387–412 (p. 407).

It did not deny subjective reason, but in some ways downgraded it by considering it merely a limited and partial expression of a universal rationality capable of providing criteria for all things and all beings. It valued ends rather than means, with a view to reconciling the objective order of the 'reasonable', as conceived by philosophy, with human existence, including self-interest and self-preservation.[139] Subjective reason as a moment within objective reason means, not only the study and use of formal logic, but also, in particular, confidence in the individual's ability to assert objective knowledge against the specific interests of a group and even against the deceitful consensus of the many.[140]

139 Horkheimer, *Zur Kritik der instrumentellen Vernunft*, p. 28; *Eclipse of Reason*, p. 3.

140 See, for example, *Eclipse of Reason*, p. 10; *Zur Kritik der instrumentellen Vernunft*, p. 33: 'Socrates [...] fought against both ideologic conservatism and relativism masked as progressiveness but actually subordinated to personal and professional interests. In other words, he fought against the subjective, formalistic reason advocated by the other Sophists'.

III.
DOMINATION OVER NATURE IN *DIALECTIC OF ENLIGHTENMENT* AND *ECLIPSE OF REASON*

1. *The ambivalence of the relationship between human beings and nature in* Dialectic of Enlightenment

In *Dialectic of Enlightenment*, the concept of nature and the question of the relationship between man and nature taking the form of domination emerge from the preface to the first edition and in the initial pages of the first chapter onwards. The preface emphasises the inextricability of subjugation to nature and social progress in today's society: economic powers (*ökonomische Mächte*), which oppress and nullify the individual, have enabled society as a whole to achieve unprecedented domination (*Gewalt*) over nature.[1] The first chapter begins with a long quotation from Bacon, in which the English thinker argues for an adequate relationship between human understanding and 'the nature of things'.[2] Bacon shows that '[w]hat human beings seek to learn from nature is how to use it to dominate wholly [*vollends zu beherrschen*] both it and human beings'.[3] It is

1 Horkheimer and Adorno, *Dialektik der Aufklärung*, p. 20; *Dialectic of Enlightenment*, p. xvii.

2 Horkheimer and Adorno, *Dialektik der Aufklärung*, pp. 25, 26; *Dialectic of Enlightenment*, pp. 1, 2.

3 Horkheimer and Adorno, *Dialektik der Aufklärung*, p. 26; *Dialectic of Enlightenment*, p. 2. The issue of domination over nature and human beings as the goal of bourgeois society is a constant theme in Horkheimer's thinking. See for example Horkheimer, 'Materialismus und Metaphysik', p. 84; 'Materialism and Metaphysics', p. 24: 'The materialism of the early bourgeois era aimed, on the contrary, at developing the knowledge of nature and attaining new powers [*Kräfte*] of mastery over nature and man [*zur Beherrschung von Natur und Menschen*]'.

therefore essential to systematically investigate the meaning of the notion of nature and the configuration of the relationship between human beings, society and the natural environment in Western civilisation, by identifying its genealogy and outcomes from the perspective of the two authors in *Dialectic of Enlightenment*.

Gunzelin Schmid Noerr notes a 'fundamental ambiguity of the concept of nature' in the *Dialectic of Enlightenment*. Nature appears as a framework of violence, as killing and devouring. This continues into society, which in this respect remains a piece of natural history: a war of all against all, as in Hobbes. Nevertheless, nature comes across as a place of living existence; its objective intention, in the Rousseauian sense, is freedom from suffering and physical happiness. Idealised images of non-violent synthesis are an expression of this intention.[4] According to Gerhard Schweppenhäuser, the intertwining of contradictory motives is described in *Dialectic of Enlightenment* as an inevitable result. This has always been the controversial point in the reception of the book. In this light, Horkheimer and Adorno's concept of nature appears not only ambivalent, but aporetic. If it is true that science always defines 'the essence of things' solely as 'the substrate of domination' and constitutes 'the unity of nature' from such an 'identity', then a definition of nature that does not degrade nature to the substrate of domination could not be considered scientific knowledge. Science would be identical with the rationality of domination, and the only alternative model would be irrationalism.[5]

As is well known, Horkheimer and Adorno criticise Western reason by showing that it is compromised by barbarism and leads

---

4 Gunzelin Schmid Noerr, *Das Eingedenken der Natur im Subjekt. Zur Dialektik von Vernunft und Natur in der Kritischen Theorie Horkheimers, Adornos und Marcuses* (Darmstadt: Wissenschaftliche Buchgesellschaft, 1990), p. ix.

5 Gerhard Schweppenhäuser, 'Nominalismuskritik und "negative Metaphysik". Philosophiegeschichtliche Überlegungen zum Begriff der Natur in der Dialektik der Aufklärung', in *Zur Kritik der regressiven Vernunft. Beiträge zur 'Dialektik der Aufklärung'* ed. by Gunzelin Schmid Noerr and Eva-Maria Ziege (Wiesbaden: Springer, 2019), pp. 65–89 (p. 69).

to self-destruction. This critique also includes questioning the unlimited domination of human reason over nature. The basic thesis of the book can be expressed as follows: the *Aufklärung*, 'understood in the widest sense as the advance of thought',[6] whose essential core throughout the centuries has been scientific and anti-mythical reason, falls back into mythology and barbarism – as shown by the taboo that positivism imposes with its pure immanence on facts, civilisation as the agent of its own destruction, the reduction of man and intelligence to commodities in the culture industry; conversely, mythology is already *Aufklärung* in itself.[7] This circle of myth and enlightenment reason has its pivot in the common opposition to nature, which in the activity of constitution and self-preservation of the ego is purely an object of intellectual inquiry and subservience to technology.[8] For Horkheimer and Adorno, the *Aufklärung*, insofar as it signifies scientific and anti-mythical reason, also designates and encompasses the actual process of increasing technical domination of nature – the (Weberian) disenchantment of the world[9] – as well as Francis Bacon's idea, according to which knowledge, as liberation from superstition,[10] is a source of power and 'is to rule over disenchanted nature [*soll über die entzauberte Natur gebieten*]'.[11]

Indeed, the problem of domination appears already in the 'Preface' of *Dialectic of Enlightenment*, in which the authors note the current

6 Max Horkheimer and Theodor W. Adorno, *Dialektik der Aufklärung*, p. 25; *Dialectic of Enlightenment*, p. 2.

7 Horkheimer and Adorno, *Dialektik der Aufklärung*, p. 30; *Dialectic of Enlightenment*, 'The Concept of Enlightenment', p. 5.

8 Furio Cerutti, 'Marxismo e sociologia nella Repubblica federale tedesca', *Belfagor*, 24.6 (1969), pp. 633–52 (p. 646).

9 'The disenchantment of the world [*Die Entzauberung der Welt*] means the extirpation of animism'. Horkheimer and Adorno, *Dialektik der Aufklärung*, p. 27; *Dialectic of Enlightenment*, p. 2.

10 'Enlightenment's program was the disenchantment of the world [*die Entzauberung der Welt*]'. Horkheimer and Adorno, *Dialektik der Aufklärung*, p. 25; *Dialectic of Enlightenment*, p. 1. This means freeing the minds of men and the world from the presence of magic and superstition.

11 Horkheimer and Adorno, *Dialektik der Aufklärung*, p. 26; *Dialectic of Enlightenment*, p. 2.

interconnection between mastery of nature and social progress. It functions ambiguously to bridge two levels of analysis.

> The enslavement to nature of people [*Die Naturverfallenheit der Menschen*] today [*heute*] cannot be separated from social progress. [...] The individual is entirely nullified in face of the economic powers [*gegenüber den ökonomischen Mächten*]. These powers are taking society's domination [*die Gewalt*] over nature to unimagined heights.[12]

Domination over nature as a response to the fear that its forces instil in human beings, manifests itself and is achieved through two interconnected modes: the levelling of qualitative differences existing in nature through processes of abstraction and mathematical formalisation, and the technical manipulation and industrial transformation of natural resources. Indeed, the subjugation of nature is based on the reduction of nature to chaotic matter that is compartmentalised and classified by the omnipotent ego.[13] A single relationship thus takes over from the variety and multiplicity of which the world of magic was a part. It is that which remains between 'the subject who confers meaning [*sinngebendem*] and the meaningless [*sinnlosem*] object'.[14] Nevertheless, the human appropriation of nature 'is not a socially innocent resource'.[15] The domination of external nature through technology and industry implies a social domination (discipline and submission) that in modern times is expressed in the forms of self-control and self-repression, in brief, in the domination of internal nature.[16] Put another way, the domination of humans over nature leads to the domination of some individuals over others –

12 Horkheimer and Adorno, *Dialektik der Aufklärung*, p. 20; *Dialectic of Enlightenment*, p. xvii.

13 See Horkheimer and Adorno, *Dialektik der Aufklärung*, p. 32; *Dialectic of Enlightenment*, p. 6.

14 Horkheimer and Adorno, *Dialektik der Aufklärung*, p. 33; *Dialectic of Enlightenment*, p. 7.

15 Simon Jarvis, *Adorno: A Critical Introduction* (New York: Routledge, 1998), p. 27.

16 On the relations of domination in the modern era, see Horkheimer, 'Egoismus und Freiheitsbewegung', pp. 17–21; 'Egoism and Freedom Movement', pp. 55–8.

indeed, the power of a minority over the multitude of workers.[17] The individual is thus completely reified: he is an object of statistics, subjected to the criteria of self-preservation (*Selbsterhaltung*) and 'successful or unsuccessful adaptation (*Angleichung*) to the objectivity of their function and the schemata assigned to it'.[18] Attempts to suspend natural constraints and coercions end up creating new constraints and coercions and subjugating human subjects themselves to the inexorability of calculation and prediction – and therefore repeatability –, based on the reduction of nature to mathematical formulas, on its conceptualisation in mathematical terms.[19] Horkheimer and Adorno revisit Galileo's 'mathematisation' of nature in the sense in which Husserl uses the term. In *The Crisis of European Sciences and Transcendental Phenomenology*, it denotes the process through which nature is turned into 'a mathematical manifold'[20] and qualitative events are indirectly mathematised through measurement.[21] Taking into account Husserl's analyses, Horkheimer and Adorno highlight how this process ends up creating

---

17 'Power [*Herrschaft*] confers increased cohesion and strength on the social whole in which it is established. The division of labor, through which power manifests itself socially, serves the self-preservation of the dominated whole [*dem beherrschten Ganzen*]'. Horkheimer and Adorno, *Dialektik der Aufklärung*, p. 44; *Dialectic of Enlightenment*, p. 16.

18 Horkheimer and Adorno, *Dialektik der Aufklärung*, p. 51; *Dialectic of Enlightenment*, pp. 21–2.

19 Horkheimer and Adorno, *Dialektik der Aufklärung*, p. 47; *Dialectic of Enlightenment*, pp. 18–19. The reduction of nature to mathematical multiplicity goes hand in hand with the reification of thought, which is reduced to an automatic process, renouncing self-reflection. Horkheimer and Adorno, *Dialektik der Aufklärung*, pp. 47–8; *Dialectic of Enlightenment*, p. 19.

20 Edmund Husserl, *The Crisis of European Sciences and Transcendental Phenomenology*, trans. by David Carr (Evanston, Illinois: Northwestern University Press, 1970), § 9, 'Galileo's mathematization of nature', p. 23. As is known, Horkheimer and Adorno refer to Husserl's book and his idea that nature became a mathematical manifold. See Horkheimer and Adorno, *Dialektik der Aufklärung*, p. 47; *Dialectic of Enlightenment*, p. 19.

21 Husserl, *The Crisis of European Sciences and Transcendental Phenomenology*, p. 38.

a crystallisation of critical thinking, a general levelling and a new type of enslavement of minds and bodies in the capitalist system.[22]

> Any attempt to break the compulsion of nature [*in den Naturzwang*] by breaking nature only succumbs more deeply to that compulsion [*den Naturzwang*]. That has been the trajectory of European civilization. Abstraction, the instrument of enlightenment, stands in the same relationship to its objects as fate, whose concept it eradicates: as liquidation. Under the levelling rule of abstraction [*Unter der nivellierenden Herrschaft des Abstrakten*], which makes everything in nature repeatable, and of industry, for which abstraction prepared the way, the liberated finally themselves become the 'herd' [*Trupp*], which Hegel identified as the outcome of enlightenment.[23]

The analysis of the itinerary and adventures of Odysseus – intended to prove the dialectic of enlightenment[24] – shows particularly how natural bonds are replaced by social ones, which are self-imposed.[25] The tale of Odysseus reveals the constitution of the Self through the multiplicity of mythical figures whom he confronts and struggles against.[26] The Self (of the bourgeois subject, prefigured by Odysseus) is constituted by estranging itself from nature, yet without completely renouncing its call. Odysseus' stratagem – listening to the song of the Sirens while remaining tied up so as not to succumb to its call – allows the hero not to be absorbed by the mythical-natural forces, but at the price of a programmatic and drastic management of his

22 See for example these passages: 'Whatever might be different is made the same. […] Not merely are qualities dissolved in thought, but human beings are forced into real conformity'. Horkheimer and Adorno, *Dialektik der Aufklärung*, pp. 34–5; *Dialectic of Enlightenment*, pp. 8–9.

23 Horkheimer and Adorno, *Dialektik der Aufklärung*, p. 35; *Dialectic of Enlightenment*, p. 9.

24 See for example: Horkheimer and Adorno, *Dialektik der Aufklärung*, pp. 57–8; *Dialectic of Enlightenment*, p. 27: 'Measures like those taken on Odysseus's ship in face of the Sirens are a prescient allegory of the dialectic of enlightenment'.

25 See Steven Vogel, *Against Nature: The Concept of Nature in Critical Theory* (Albany: State University of New York Press, 1996), p. 53.

26 Horkheimer and Adorno, *Dialektik der Aufklärung*, p. 71; *Dialectic of Enlightenment*, p. 38.

body, a rigorous discipline that allows him to curb and keep his impulses at bay.

> Odysseus, like the heroes of all true novels after him, throws himself away, so to speak, in order to win himself; he achieves his estrangement from nature [*die Entfremudng von der Natur*] by abandoning himself to nature [*in der Preisagbe an die Natur*], trying his strength against it in all his adventures; ironically, it is implacable nature that he now commands, which triumphs on his return home as the implacable judge, avenging the heritage of the very powers he has escaped. At the Homeric stage, the identity of the self is so much a function of the nonidentical [*die Identität des Selbst Funktion des Unidentischen*], of dissociated, unarticulated myths, that it must derive itself from them.[27]

In the *Dialectic of Enlightenment*, Odysseus' ruse or μηχανή symbolizes an archetypal action in the process of 'self-formation' just as the circumvention of the Sirens' island captures a key moment in the genesis of rational subjectivity.[28] Within this process the relationship between Odysseus and nature remains dialectical: the bourgeois and rational Self, the Self that recognises itself as identical in the multiplicity of its actions and thoughts and in its managing to survive mortal dangers,[29] emerges and is constituted by entering into a relationship with a myriad of mythical figures, 'natural deities [*Naturgottheiten*]', which it manages to confront – in such a way as to preserve itself and not perish – with cunning (*die List*).[30] Furthermore, as noted in the passage quoted above, nature is never completely dominated and silenced.[31] It re-emerges – in the form of

27 Horkheimer and Adorno, *Dialektik der Aufklärung*, p. 72; *Dialectic of Enlightenment*, pp. 38–9.

28 On this point, see Felix Kämper, 'Domination by nature: On an unresolved tension in the Frankfurt School', *Thesis Eleven*, 2025, 1–16 (p. 3).

29 Horkheimer and Adorno, *Dialektik der Aufklärung*, p. 55; *Dialectic of Enlightenment*, p. 25.

30 Horkheimer and Adorno, *Dialektik der Aufklärung*, p. 78; *Dialectic of Enlightenment*, p. 42.

31 In this regard, consider also the observations made by Adorno in his study of Kierkegaard, in the pages devoted to the hostility that the sovereign spirit has always shown towards nature. According to Adorno, the separation of spirit from nature and the body result in their revenge on

uncontrollable primordial impulses – when he returns home and takes ruthless revenge on the usurpers of his home, renouncing the law and the rules of society. Indeed, in 'The Concept of Enlightenment', Horkheimer and Adorno show how enlightenment fails to eliminate all traces of nature: its voice is 'audible in its estrangement', 'something blind and mutilated [*als Blindesm Verstümmeltes*]'.[32] Abandoning oneself to nature (*Naturverfallenheit*) consists in subjugating nature (*besteht in der Naturbeherrschung*), through which the mind is formed and exists. When the mind (*der Geist*) recognises itself as separate from nature (more precisely as nature split from itself), identifies its action as dominion (*Herrschaft*) and is taken back into nature, then the claim to dominion (*herrschaftliche Ansprung*) that enslaves it to nature, also dissolves.[33]

Consider now the following passage, taken from the 'Excursus I: Odysseus or Myth'.

> In class society, the self's hostility to sacrifice included a sacrifice of the self, since it was paid for by a denial of nature in the human being for the sake of mastery over extrahuman nature and over other human beings [*der Herrschaft über die außermenschliche Natur und über Andere Menschen*]. This very denial, the core of all civilizing rationality, is the germ cell of proliferating mythical irrationality: with the denial of nature in human beings, not only the *telos* of the external mastery of nature [*das Telos der auswendigen Naturbeherrschung*] but also the *telos* of one's own life becomes confused and opaque. At the moment when human beings cut

---

spirit to the extent that spirit cannot escape from what it aims to subjugate. See Theodor W. Adorno, *Kierkegaard: Construction of the Aesthetic*, trans. by Robert Hullot-Kentor (Minneapolis: University of Minnesota Press, 1989), pp. 52–53, 104–5. In this text, on the one hand, 'nature' refers to specific theological questions and an anthropology understood in the traditional theological sense; on the other, it designates the opposite of the spirit, what is not spiritual. On this point, see Roland Boer, 'A Totality of Ruins: Adorno on Kierkegaard', *Cultural Critique*, 83 (2013), 1-30 (p. 17).

32 Horkheimer and Adorno, *Dialektik der Aufklärung*, p. 63; *Dialectic of Enlightenment*, p. 31.

33 Horkheimer and Adorno, *Dialektik der Aufklärung*, p. 63; *Dialectic of Enlightenment*, p. 31.

> themselves off from the consciousness of themselves as nature, all the purposes for which they keep themselves alive – social progress, the heightening of material and intellectual forces, indeed, consciousness itself – become void, and the enthronement of the means as the end, which in late capitalism is taking on the character of overt madness, is already detectable in the earliest history of subjectivity. The human being's mastery of itself, on which the self is founded, practically always involves the annihilation of the subject in whose service that mastery is maintained, because the substance which is mastered, suppressed, and disintegrated by self-preservation is nothing other than the living entity, of which the achievements of self-preservation can only be defined as functions – in other words, self-preservation destroys the very thing which is to be preserved.[34]

With the denial and repression of the living nature present in human beings, the Self loses sight of the ends and meaning of the very process of subjugating nature (*Naturbeherrschung*), which results in its global domination (*der Herrschaft über die außermenschliche Natur und über Andere Menschen*). Where everything should be clearer and more distinct – thanks to modern science and the mathematisation of nature –, social reality and individuals' lives instead become confused and opaque.

The relationship between the human subject and nature, the identical and unified Self and the non-identical plurality of objects, is the key to understanding the peculiarity of the Enlightenment. In order to overcome the fear of nature, to free it from magical powers and myth, the Enlightenment seeks to 'disenchant' the world, but ends up proposing new enchantments, thus falling back into mythology.[35] Indeed, the 'disenchantment of the world' comes

34 Horkheimer and Adorno, *Dialektik der Aufklärung*, p. 78; *Dialectic of Enlightenment*, pp. 42–3.

35 On the connection between mythical figures and fear of the power of nature, see in particular the following passage: 'Enlightenment has always regarded anthropomorphism, the projection of subjective properties onto nature, as the basis of myth. The supernatural, spirits and demons, are taken to be reflections of human beings who allow themselves to be frightened by natural phenomena'. Horkheimer and Adorno, *Dialektik der Aufklärung*, pp. 28–9; *Dialectic of Enlightenment*, p. 4.

about through the denial of the internal nature of human beings: it allows them to control the external nature. In its most original dimension, from a genealogical perspective, as demonstrated above, this process is nothing more than the development of the ability to suppress instincts, which makes human beings into selves (subjects), and distinguishes them from animals. Becoming a Self, a rational subject, as shown by the entire story of Odysseus and, specifically, the references to men turned into pigs in the episode of Circe, means freeing oneself from the grip of nature.[36] Nevertheless, this process, on the one hand, results in new forms of mythical irrationality, on the other hand, in new system of subjugation of human beings. Therefore, while the enlightenment seeks to free man from slavery and domination, in reality, it reproduces new forms of slavery and strengthens domination over man. As Lucio Cortella points out, this paradoxical nature of the enlightenment can be explained by looking at the very origins of the enlightenment process, that is, the roots of human civilisation. This began with a very specific act: the separation of human beings from nature, their differentiation and opposition to it.[37] The dialectic and paradox of enlightenment originate from this opaque relationship between the Self (the bourgeois, rational subject) and nature, which ultimately leads to a short circuit: in human beings' mastery (*Herrschaft*) of themselves – the foundation of their dominion over external nature – what is dominated and oppressed by the Self that wants to preserve itself, is nothing other than *das Lebendige*, the living being (since the natural component to be managed and repressed is his own bodily life), from which, and in relation to which, the methods

36 'The suppression of instinct which constitutes them as selves and separates them from beasts was the introverted form of the repression existing within the hopelessly closed cycle of narure, to which, according to an earlier theory, the name Circe alludes'. Horkheimer and Adorno, *Dialektik der Aufklärung*, p. 94; *Dialectic of Enlightenment*, p. 55.

37 On this point, see the brilliant explanation of Lucio Cortella, *Per una dialettica nella finitezza. Adorno e il programma di una dialettica negativa* (Rome: Meltemi, 2006), p. 7.

and strategies of self-preservation are defined and organised.[38] For this reason, paradoxically, the domination of human beings over themselves is virtually the destruction of the subject in whose service it is realised. In reality, here Horkheimer and Adorno present the paradox with a synecdoche: the domination of some individuals over others – that allows the Self to be constituted – involves the repression of the fundamental and irreducible natural, living component of the subject, which – precisely by synecdoche – means the destruction of the subject itself. This paradox evokes the process described by Nietzsche, whereby human beings forget their animal origins and natural roots to the point of developing a sense of shame about their instincts and considering life itself to be 'distasteful'.[39]

---

38 See Horkheimer and Adorno, *Dialektik der Aufklärung*, p. 78; *Dialectic of Enlightenment*, p. 43.

39 'The tired, pessimistic outlook, mistrust of life's riddle, the icy 'no' of nausea at life – these are not signs of the wickedest epoch of the human race: on the contrary, they come to light as the bog-plants they are only in their natural habitat, the bog, – I mean the sickly mollycoddling and sermonizing, by means of which the animal 'man' is finally taught to be ashamed of all his instincts. On the way to becoming an 'angel' (not to use a stronger word here), man has upset his stomach and developed a furry tongue so that he finds not only that the joy and innocence of animals is disgusting, but that life itself is distasteful [...]'. Friedrich Nietzsche, *On the Genealogy of Morality*, ed. by Keith Ansell-Pearson, trans. by Carol Diethe (Cambridge and New York: Cambridge University Press, 2006), 'Second essay', § 7, p. 43. As is well known, Nietzsche, along with Kant and Sade, is a key author in the 'Excursus II: Juliette or Enlightenment and Morality' of *Dialectic of Enlightenment*. Here, speaking of Christianity's attempt to reconcile nature and civilisation, Horkheimer and Adorno present Nietzsche as the philosopher who, opposing monotheism and Christianity in particular, believes it is possible to overcome the dissolution of bonds and values through the figure of the *Übermensch*, the 'higher self'. Horkheimer and Adorno, *Dialektik der Aufklärung*, pp. 137–8; *Dialectic of Enlightenment*, pp. 89–90. On Nietzsche, Horkheimer and Adorno, see Vanessa Lemm, 'Critical theory and affirmative biopolitics: Nietzsche and the domination of nature in Adorno/Horkheimer', *Journal of Power*, 3.1 (2010), 75–95. See in particular her description of Nietzsche's antagonism between civilisation and culture: the former 'represents the perspective of error that turns

In summary, the Self (the subject, the reason) that aims for self-preservation, denies nature and the living body in which it is incorporated and ends up destroying itself. In the light of the passages already examined and of other fundamental texts by Horkheimer of the 1940s ('The End of Reason' and *Eclipse of Reason*), it can be stated that this self-destruction manifests itself on three levels:

1) the invalidation of the fundamental concepts and values of bourgeois society (ultimately, the very notion of reason);[40]

2) the inconsistencies and contradictions of the capitalist system which, despite its epistemological and technological potential, fails to meet the needs and resolve the difficulties of a significant part of the population;[41]

---

animals into human', the latter 'represents the persepctive of truth that turns humans back into animals'. Lemm, 'Critical theory and affirmative biopolitics', p. 78. Moreover, on this point, consider Habermas's interpretation: according to him, '*Genealogy of Morals* was the great model for the *Dialectic of Enlightenment*'. In Nietzsche, Horkheimer and Adorno find 'the two elements of the domination over external and internal nature' which 'combine and solidify into the institutionalized domination of men over men'. Jürgen Habermas, 'The Entwinement of Myth and Enlightenment: Re-Reading Dialectic of Enlightenment', trans. by Thomas Y. Levin, *New German Critique*, 26 (Spring–Summer, 1982), *Critical Theory and Modernity*, 13–30 (pp. 22 and 24, respectively).

40 Max Horkheimer, 'Vernunft und Selbsterhaltung', in Horkheimer, *GS* 5, pp. 320–50 (pp. 320–2); 'The End of Reason', *Studies in Philosophy and Social Science*, 9:3 (1941), 366–88 (pp. 366–7). See also Max Horkheimer, *Eclipse of Reason* (New York: Oxford University Press, 1947), pp. 23–4; *Zur Kritik der instrumentellen Vernunft*, trans. by Aldred Schmidt, in Horkheimer, *GS* 6, pp. 19–186 (pp. 44–5).

41 In bourgeois society, the specific dynamics of the economy mean that some accumulate enormous power ('a fantastic degree of power') while others (the majority of individuals) find themselves in a state of material and intellectual powerlessness ('material and intellectual weakness'). As Horkheimer writes, '[t]he original fruitfulness of the bourgeois organization of the life process is thus transformed into a paralyzing barrenness, and men by their own toil perpetuate a reality which enslaves them to ever greater degree'. Horkheimer, 'Traditionelle und kritische Theorie', p. 186; 'Traditional and Critical Theory', pp. 212–3 (translation amended).

3) in an increasingly disenchanted world,[42] the reversal of emancipation and freedom into servitude and standardisation through an effective system of rationalisation and administration of reality as a whole results in a totalitarian system of exploitation of humans and natural resources to the point of rational administration of suffering and death in the Nazi concentration camps.[43]

The bourgeois ban on relying directly on life without a rational modulation of self-preservation ('without any rational reference to self-preservation') rejects nature, which is true self-preservation, but in doing so it creates the conditions for the unleashing thereof 'in the individual as in the collective fate of crisis and war'.[44] In relation to the extreme developments of the denial of nature and living entities in the twentieth century, Horkheimer and Adorno bring to light 'the antireason [*Widervernuft*] of totalitarian capitalism' that 'tends toward the extermination of humanity'.[45] In this way, they reintroduce into the specific context of late capitalism as the extreme configuration of the totalitarian character of Enlightenment,[46] the possibility of humans destroying each other through the control of nature, as Freud had mentioned at the end of *Civilisation and Its Discontents*.[47]

---

42 In this world, subjective-instrumental reason prevails and eclipses objective reason, i.e., that form of reason which identifies ultimate and general purposes in the world.

43 See Horkheimer and Adorno, *Dialektik der Aufklärung*, pp. 42–3; *Dialectic of Enlightenment*, p. 15.

44 Horkheimer and Adorno, *Dialektik der Aufklärung*, pp. 52 and 53; *Dialectic of Enlightenment*, pp. 22 and 23.

45 Horkheimer and Adorno, *Dialektik der Aufklärung*, p. 78; *Dialectic of Enlightenment*, p. 43.

46 See Horkheimer and Adorno, *Dialektik der Aufklärung*, p. 47; *Dialectic of Enlightenment*, p. 18.

47 'The fateful question for the human species seems to me to be whether and to what extent their cultural development will succeed in mastering the disturbance in their communal life by the human instinct of aggression and self-destruction. It may be that in this respect precisely the present time deserves a special interest. Men have gained control over the forces of nature to such an extent that with their help they would have no difficulty in exterminating one another to the last man'. Sigmund Freud, *Civilization and Its Discontents* (1929), in The Standard Edition of the Complete Psychological Works of Sigmund Freud, translated from the

This being the case, as we have shown by examining one of the last pages of the chapter 'The Concept of Enlightenment', only reason that takes into account the ambivalence in the relationship between civilisation and nature and the dialectical character of the domination over nature (which reverses itself in domination by nature),[48] only a Self that understands itself as nature in conflict with itself, can – through the 'remembrance [*Eingedenken*] of nature within the subject'[49] – escape the modern destiny of the *Aufklärung*, without the ego losing its identity, but rather reconciling itself with nature beyond repression and domination.[50] Conversely, while the process which tends to extirpate nature, idealising it on the model of mathematics, also produces the reification of thought,[51] the ability

German under the general editorship of James Strachey, in collaboration with Anna Freud; assisted by Alix Strachey and Alan Tyson, Volume xxi (1927–1931): *The Future of an Illusion, Civilization and its Discontents and Other Works* (London: The Hogarth Press and Institute of Psycho-Analysis, 1961), pp. 64–145 (p. 145).

48 'In the mastery of nature, without which mind does not exist, enslavement to nature persists'. Horkheimer and Adorno, *Dialektik der Aufklärung*, p. 63; *Dialectic of Enlightenment*, p. 31. Consider again the episode of the Sirens: Odysseus understands that, however much he may distance himself from nature, he remains, as a listener, enslaved to it ('*ihr verfallen bleibt*'). Ibid., p. 83; English trans., p. 46. According to Vogel, the Odyssean cunning itself consists of 'the master of nature through [...] adaptation [to nature]. [...] The curious outcome of the project of dominating nature by obeying it then is that *nature wins* – and that its victory turns out to have been implicit in and even presupposed from the very beginning'. Vogel, *Against Nature*, p. 55.

49 Horkheimer and Adorno, *Dialektik der Aufklärung*, p. 64; *Dialectic of Enlightenment*, p. 32.

50 See Cerutti, 'Marxismo e sociologia nella Repubblica federale tedesca', pp. 646–7.

51 'Thought is reified as an autonomous, automatic process, aping the machine it has itself produced, so that it can finally be replaced by the machine'. Horkheimer and Adorno, *Dialektik der Aufklärung*, pp. 47–8; *Dialectic of Enlightenment*, p. 19. 'By sacrificing thought, which in its reified form as mathematics, machinery, organization, avenges itself on a humanity forgetful of it, enlightenment forfeited its own realization'. Horkheimer and Adorno, *Dialektik der Aufklärung*, p. 65; *Dialectic of Enlightenment*, p. 33.

of thought to recover 'the element of reflection on itself'[52] – which means developing an awareness of the regressive moment of the Enlightenment[53] – goes hand in hand with a possible change in the relationship between human beings and a nature conceived and practised under the sign of an ambivalent domination.

2. *Domination and mutilation: Genealogy of the mastery over nature and its outcome*

For Horkheimer and Adorno, despite the development of their position on this subject over the years, the origin of all 'hegemonic humanism' and its misunderstandings, is already inscribed, not only in the Homeric narrative of the Odyssey and in the deeds of its cunning Odysseus, but also in the first account of *Genesis*, which provided the basis and justification for the postulate, now accepted by Judeo-Christian culture and European history, that the human being, as a rational being gifted with speech, can subjugate nature and animals.

The reference to *Genesis* appears in a key passage of *Dialectic of Enlightenment*. The two authors expose the thesis that bourgeois society is governed by the principle of equivalence ('*ist beherrscht vom Äquivalent*'), through which dissimilar things became comparable to the extent that they are reduced to abstract quantities.[54] The entire philosophical history, from Parmenides to Russell, is obsessed by the idea of 'unity' (*Einheit*) and by the effort to destroy 'all gods and qualities'.[55]

The equivalence is not only a logical-philosophical principle: concretely it is the principle ruling the capitalistic world. We should

52 Horkheimer and Adorno, *Dialektik der Aufklärung*, p. 60; *Dialectic of Enlightenment*, p. 29.

53 Horkheimer and Adorno, *Dialektik der Aufklärung*, pp. 18–19; *Dialectic of Enlightenment*, p. xvi.

54 Horkheimer and Adorno, *Dialektik der Aufklärung*, p. 30; *Dialectic of Enlightenment*, p. 4.

55 Horkheimer and Adorno, *Dialektik der Aufklärung*, p. 30; *Dialectic of Enlightenment*, p. 5.

bear in mind that equivalence is '[t]he secret of the commodity form, which Marx described in the first chapter of *Capital*: the equivalence of one object or activity with a heterogeneous other. Equating unequal or non-identical things, exchange is the social model for what Adorno called identity thinking: a form of concept fetishism in which objects are summarily identified with concepts'.[56] Generalised equivalence entails the de-substantialisation of the concept of nature. Horkheimer and Adorno attribute it, from a social-philosophical perspective, to the social conditions under which scientific and technological domination of nature frees itself from its metaphysical constraints.[57] These conditions create a universal mediation of coexistence through commodities.[58] Nevertheless, the compulsion to generalise inherent in scientific methods is conducive to an alliance with the rationality of production and administration, but it does not necessarily lead to it. It can become part of it, but it does not inevitably produce it. The logical application of scientific methods

---

56 Cook, *Adorno on Nature*, p. 19. See Adorno, *Negative Dialektik*, pp. 149–50; *Negative Dialectics*, p. 146: 'If mankind [*die Menschheit*] is to get rid of the coercion to which the form of identification really subjects it, it must attain identity with its concept at the same time. In this, all relevant categories play a part. The barter principle, the reduction of human labor to the abstract universal concept of average working hours, is fundamentally akin to the principle of identification. Barter [*Tausch*] is the social model of the principle, and without the principle there would be no barter; it is through barter that nonidentical individuals and performances become commensurable and identical. The spread of the principle imposes on the whole world an obligation to become identical, to become total [*Die Ausbreitung des Prinzips verhält die ganze Welt zum Identischen, zur Totlität*]. But if we denied the principle abstractly – if we proclaimed, to the greater glory of the irreducibly qualitative, that parity should no longer be the ideal rule – we would be creating excuses for recidivism into ancient injustice. From olden times, the main characteristic of the exchange of equivalents has been that unequal things would be exchanged in its name, that the surplus value of labor would be appropriated'.

57 Schweppenhäuser, 'Nominalismuskritik und "negative Metaphysik"', p. 70.

58 Horkheimer, 'Vernunft und Selbsterhaltung', p. 339; 'The End of Reason', p. 379.

to isolate and control the reproduction of natural causal processes, as well as their representation in the form of natural laws, does not in itself lead to the de-substantialisation of the concept of nature.[59] This de-substantialisation depends on a complex and diachronic process in which *Aufklärung* destroys any form or track of mythology (suspected of dangerous collusion with the forces of nature that they are supposed to neutralise) just as it seeks to master nature through the alliance between science, technology and bureaucracy.

This obsession of mastery through equivalence goes back to the dawn of Western civilisation and to its Asiatic origins: the Bible and the Greek poems.

> The Olympian deities are no longer directly identical with elements, but signify them. In Homer Zeus controls the daytime sky, Apollo guides the sun; Helios and Eos are already passing over into allegory. The gods detach themselves from substances to become their quintessence. From now on, being is split between logos – which, with the advance of philosophy, contracts to a monad, a mere reference point – and the mass of things and creatures in the external world. The single distinction between man's own existence and reality swallows up all others. Without regard for differences, the world is made subject to man. In this the Jewish story of creation and the Olympian religion are at one: '… and let them have dominion over the fish of the sea, and over the fowl of the air, and over the cattle, and over all the earth, and over every creeping thing that creepeth upon the earth'. [Genesis, I, 26] 'O Zeus, Father Zeus, yours is the dominion of the heavens; you oversee the works of men, both the wicked and the just, and the unruly animals, you who uphold righteousness'.[60]

Here we find the origin of the immemorial 'separation of spirit from physis, the separation of the labour of the mind from that of the hands', as Adorno writes in *Philosophy of New Music*: as Hegel's dialectic of master and slave illustrates, 'the more [the] mind [that dominates nature] advances toward autonomy, the more it distances itself from

59 Schweppenhäuser, 'Nominalismuskritik und "negative Metaphysik', p. 70.

60 Horkheimer and Adorno, *Dialektik der Aufklärung*, pp. 30–1; *Dialectic of Enlightenment*, p. 5.

a concrete relationship to all it dominates, men and materials alike'.[61] In the process of rationalisation, mathematisation and disenchantment through which the world is made subject to humans beings, Horkheimer and Adorno detect a rupture between subject and object, man and nature, a split through which a natural being forgets its natural origin, turning against and mutilating it:[62] indeed, as the agency of calculating thought, reason organises the world for the purposes of self-preservation and simply shapes and manipulates the object to turn it into mere material content to be subjugated and exploited.[63] In this vein, Horkheimer and Adorno write:

> The true nature of the schematism which externally coordinates the universal and the particular, the concept and the individual case, finally turns out, in current science, to be the interest of industrial society. Being is apprehended in terms of manipulation and administration. Everything – including the individual human being, not to mention the animal – becomes a repeatable, replaceable process, a mere example of the conceptual models of the system.[64]

Nevertheless, Horkheimer and Adorno strongly emphasise humanity's roots in nature and animality. Even what is called 'progress', which seems specifically human in origin, happens 'when human beings become conscious of their own naturalness and call a halt to their own domination of nature', insofar as they understand that progress is also nature and that through such a domination 'nature's own domination is perpetuated'.[65] They

---

61 Theodor W. Adorno, *Philosophie der neuen Musik*, *GS* 12 (Frankfurt am Main: Suhrkamp, 2003); *Philosophy of New Music*, trans, edited, and with an introduction by Robert Hullot-Kentor (Minneapolis and London: University of Minnesota Press, 2006), p. 20.

62 Vuillerod, *Theodor W. Adorno. La domination de la nature*, pp. 68–9.

63 Horkheimer and Adorno, *Dialektik der Aufklärung*, p. 106; *Dialectic of Enlightenment*, p. 65.

64 Horkheimer and Adorno, *Dialektik der Aufklärung*, pp. 106–7; *Dialectic of Enlightenment*, p. 65.

65 Adorno, *Zur Lehre von der Geschichte und von der Freiheit*, p. 214; *History and Freedom*, pp. 151–2. See also Theodor W. Adorno, *Stichworte. Kritische Modelle 2*, in Adorno, *GS* 10.2: *Kulturkritik und Gesellschaft II*: *Eingriffe – Stichworte – Anhang*, 7th edition (Frankfurt am Main:

mark the continuity between humans and nature by inscribing the tendency towards domination in nature itself. The desire to dominate the world through reason can be explained by the natural instinct to preserve one's own life against the hostile forces of external and internal nature. As Horkheimer and Adorno state in the 'Notes and Sketches' section of *Dialectic of Enlightenment*, '[o]nly the cunning strength [*die abgefeimte Kraft*] which survives is in the right. But that strength itself is only nature; the whole ingenious machinery [*die ganze ausgetüftelte Maschinerie*] of modern industrial society is no more than nature dismembering itself'.[66] Although enlightenment poses 'humanism as its guiding value', without managing to elucidate 'why that value (or any) value should guide it',[67] human domination over nature is understood from the tendencies of nature itself, so that anthropocentrism is profoundly challenged.[68] While, by a strange paradox, domination over nature is generated by nature itself, it seeks to neutralise its subjugating forces and, finally, 'is paid for with the naturalization of social domination'.[69] Indeed, mastery over nature – which should be the real progress of human freedom –, is paid for with impotent subjection to the social divisions as well as the generalised administration and domination which grant that power. In this machinery, maintained in order to continually recreate and perpetuate domination over non-human nature and over human beings, even the self is sacrificed. At the end – as we have already seen – not only the goal of the external mastery of nature but also the end of each individual life 'becomes confused and opaque'.[70]

Suhrkamp, 2018), 'Fortschritt', pp. 617–638 (p. 625); *Critical Models: Interventions and Catchwords*, trans. by Henry W. Pickford (New York: Columbia University Press, 2005), 'Catchwords: Critical Models 2', 'Progress', pp. 143–160 (p. 150).

66 Horkheimer and Adorno, *Dialektik der Aufklärung*, pp. 285–6; *Dialectic of Enlightenment*, p. 210 (translation amended).

67 Vogel, *Against Nature*, p. 56.

68 Vuillerod, *Theodor W. Adorno. La domination de la nature*, pp. 69–70.

69 Jarvis, *Adorno: A Critical Introduction*, p. 27.

70 Horkheimer and Adorno, *Dialektik der Aufklärung*, p. 78; *Dialectic of Enlightenment*, p. 42. In *Mimima moralia*, Adorno would emphasise the disastrous consequences of domination over nature in modern bourgeois society: 'Panic breaks once again, after millennia of enlightenment, over a

This tendency towards domination has led humans in Western societies to emphasise, if not exalt, their own uniqueness, which has ruled out any attempt to consider themselves as beings *among* other living beings and in a relationship of empathy with them. In the fragment entitled 'Man and Beast', in the 'Notes and Sketches' section of *Dialectic of Enlightenment*, on the one hand, Horkheimer and Adorno highlight the aristocratic tendency of human beings to consider their condition nobler and superior to that of the animal (that 'is without concepts' and 'has no self'[71]); on the other, they focus on the result of this attitude – connected to the process of disenchantment of the world, reduced to a purely manageable and predictable quantity – namely, the estrangement of human beings from the natural element. Consider the following two passages from this section. The first extract is as follows:

> Throughout European history the idea of the human being has been expressed in contradistinction to the animal. The latter's lack of reason is the proof of human dignity. So insistently and unanimously has this antithesis been recited by all the earliest precursors of bourgeois thought, the ancient Jews, the Stoics, and the Early Fathers, and then through the Middle Ages to modern times, that few other ideas are so fundamental to Western anthropology.[72]

This passage is particularly thought-provoking and rich in meaning. According to Horkheimer and Adorno, not only does the non-human animal provide 'the material stuff of civilization' – i.e. the flesh and bone, labour and intelligence humans exploit for their purposes – but it is 'the psycho-semiotic medium' through which the entirety of human culture, philosophy and cosmology builds itself.[73] However,

---

humanity whose control of nature [*Herrschaft über Natur*] as control of men far exceeds in horror anything men ever had to fear from nature'. Adorno, *Minima moralia*, § 151, p. 273; *Minima moralia*, English trans., p. 239.

71 Horkheimer and Adorno, *Dialektik der Aufklärung*, pp. 278–9; *Dialectic of Enlightenment*, p. 205.

72 Horkheimer and Adorno, *Dialektik der Aufklärung*, p. 277; *Dialectic of Enlightenment*, pp. 203–4.

73 This expression is by John Sanbonmatsu, 'Introduction', in *Critical Theory and Animal Liberation*, ed. by John Sanbonmatsu (Plymouth:

it should be noted that here Horkheimer and Adorno clearly refer to European cultural tradition; thus their text also overshadows the idea that other ways of interacting and relating to non-human animals are possible in other civilisations. In fact, according to Horkheimer ('The End of Reason'), the self-positioning of (European) humans as rational beings – whereby reason is the faculty that separates them from non-human animals – is the way in which they free themselves from the shackles of nature.[74] On the other hand, in a typescript of 1944, posthumously published, entitled 'Contra Paulum', a draft for a critique of Paul Tillich's 'Man and Society in Religious Socialism' (1943), Adorno questions Tillich's distinguishing of human beings from those above and those below, angels and animals. The principle that fixes human beings in this definition, according to Adorno, is the same that separates them from nature, producing endless repression. Human dignity and disrespect for human beings end up coinciding. Indeed, Adorno concludes, if humans no longer emphasised their difference from animals, they might also cease to humiliate themselves in the presence of angels.[75] On the other hand, in fragments written in the same period (June 1941–April 1943), contesting any form of the hypostatisation of human nature, Adorno poses the question of the crisis of the individual in terms

Rowman & Littlefield, 2011), pp. 1–32 (p. 9).

74 'The age-old definition of reason in terms of self-preservation already implied the curtailment of reason itself. The propositions of idealistic philosophy that reason distinguishes man from the animal (propositions in which the animal is humiliated just as man is in the converse propositions of the materialist doctors) contain the truth that through reason man frees himself of the fetters of nature. This liberation, however, does not entitle man to dominate nature (as the philosophers held) but to comprehend it'. Horkheimer, 'Vernunft und Selbsterhaltung', pp. 348–9; 'The End of Reason', p. 387.

75 Theodor W. Adorno, 'Contra Paulum', in Adorno, *Briefe und Briefwechsel*, Band II: 1938-1944, ed. by Christoph Gödde und Henri Lonitz (Frankfurt am Main: Suhrkamp, 2004), pp. 475–502 (p. 483). See Paul Tillich, 'Man and Society in Religious Socialism' (1943), in Paul Tillich, *Main Works / Hauptwerke*, Volume 3 / Band 3, *Writings in the Social Philosophy and Ethics / Sozialphilosophische und ethische Schriften*, ed. by Carl H. Ratschow (Berlin, Boston: De Gruyter, 1998), pp. 487–502.

of the individual's recent changes within the capitalist relations of production and reproduction, and belies the new anthropology by showing that it is misleading and dangerous. He thinks that human beings must follow nature's example but, in this context, the concept of nature has an evident biological significance. In brief, Darwin's theory of the struggle for existence is the underlying theme of this concept and of the new anthropology, which chains human beings to an even blinder destiny.[76]

The second relevant passage highlights a consequence of the antithesis diagnosed in the previous one. It focuses on the (archaic) terror of slipping into the condition of animality as expressed in beliefs about, and stories of, the metamorphosis of human beings into animals – a terror which could evoke an error: the original error which – according to Nietzsche – transformed animals into humans.[77]

> In popular fairy tales the metamorphosis of humans into animals is a recurring punishment. To be imprisoned in an animal body is regarded as damnation. To children and peoples, the idea of such transformations is immediately comprehensible and familiar. Believers in the transmigration of souls in the earliest cultures saw the animal form as punishment and torment. The mute wildness in the animal's gaze bears witness to the horror which is feared by humans in such metamorphoses. Every animal recalls to them an immense misfortune which took place in primeval times.[78]

---

76 See Theodor W. Adorno, 'Individuum und Gesellschaft. Entwürfe und Skizzen', in *Frankfurter Adorno Blätter VIII*, ed. by Rolf Tiedemann (Munich: text + kritik, 2003), pp. 60–94. On this topic, see Massimiliano Tomba, 'Adorno's Account of the Anthropological Crisis and the New Type of Human', in *(Mis)readings of Marx in Continental Philosophy*, ed. by Jernej Habjan and Jessica Whyte (London: Palgrave Macmillan, 2014), pp. 34–50.

77 '*Truth as Circe*. – Error has transformed animals into men; is truth perhaps capable of changing man back into an animal?' Friedrich Nietzsche, *Human, All Too Human*, trans. by R. J. Hollingdale, with an introduction by Richard Schacht (Cambridge: Cambridge University Press, 1996), § 519, p. 182.

78 Horkheimer and Adorno, *Dialektik der Aufklärung*, p. 279; *Dialectic of Enlightenment*, pp. 205–6.

The paradoxical estrangement of humans from nature is expressed in the horror that humans manifest towards the possibility of a mixture of humanity and animality (as can be inferred from the aforementioned passage) and the inexorable fate of animals,[79] as well as in the idea that it is futile to show solicitude for the unreasoning animal. Indeed, it is not by chance that Western civilisation has left such a care (*Sorge*) to women.[80] Conversely, the pity expressed by fascists for animals and nature (as well as for children) should not deceive us: it conceals and reveals their desire for persecution and the nature of totalitarian rule. The caress on the child's hair or on the animal's skin reminds us that that same hand can destroy and, therefore, that all beings are equal before power: everyone is the same, no one has a specific essence.[81] Such reflections show, in a different light, how the paradoxical relationship between man and nature in late capitalism goes hand in hand with the domination of humans over other humans and with the establishment of a gender hierarchy within human societies:

> Human beings are so radically estranged from themselves and from nature that they know only how to use and harm each other. Each is merely a factor, the subject or object of some praxis, something to be reckoned with or discounted.
>
> In this world liberated from appearance – in which human beings, having forfeited reflection, have become once more the cleverest animals, which subjugate the rest of the universe when they happen not to be tearing themselves apart – to show concern for animals is considered no longer merely sentimental but a betrayal of progress.[82]

---

79 '[T]he animal's lack of reason holds it eternally captive in its form'. Horkheimer and Adorno, *Dialektik der Aufklärung*, pp. 279–80; *Dialectic of Enlightenment*, p. 206.

80 Horkheimer and Adorno, *Dialektik der Aufklärung*, p. 280; *Dialectic of Enlightenment*, p. 206.

81 Horkheimer and Adorno, *Dialektik der Aufklärung*, p. 285; *Dialectic of Enlightenment*, p. 210.

82 Horkheimer and Adorno, *Dialektik der Aufklärung*, p. 286; *Dialectic of Enlightenment*, p. 211.

Horkheimer and Adorno do not thematise nature as a totality, a unified and homogenic kingdom, and yet in certain passages they seem to talk about the domination and mutilation of nature in general.

> Enlightenment is more than enlightenment, it is nature made audible in its estrangement [*in ihrer Entfremdung*]. In mind's self-recognition as nature divided from itself, nature, as in prehistory, is calling to itself, but no longer directly by its supposed name, which, in the guise of *mana*, means omnipotence, but as something blind and mutilated [*sondern als Blindes, Verstümmeltes*]. In the mastery of nature [*in der Naturbeherrschung*], without which mind does not exist, enslavement to nature [*Naturverfallenheit*] persists. By modestly confessing itself to be power and thus being taken back into nature, mind rids itself of the very claim to mastery which had enslaved it to nature.[83]

Paradoxically, despite being silent in the modern era of the mathematisation of the world and the reduction of qualitative differences to mere quantities, nature makes itself felt precisely in its alienation, in its absence.

Conversely, the mutilation of nature is also the mutilation of the individual. When humans mutilate nature by dominating natural beings, that is reducing them to quantities, they also injure themselves.

> On the way from mythology to logistics [*von der Mythologie zur Logistik*], thought has lost the element of reflection on itself, and machinery mutilates people today [*die Machinerie verstümmelt die Menschen heute*], even if it also feeds them. In the form of machines, however, alienated reason [*entfremdete Ratio*] is moving toward a society which reconciles thought, in its solidification as an apparatus both material and intellectual, with a liberated living element, and relates it to society itself as its true subject.[84]

This passage is part of a broader discussion on the reduction of thought to the production of uniformity. As demonstrated above,

83 Horkheimer and Adorno, *Dialektik der Aufklärung*, p. 63; *Dialectic of Enlightenment*, p. 31.

84 *Dialektik der Aufklärung*, p. 60; *Dialectic of Enlightenment*, p. 29 (translation amended).

according to Horkheimer and Adorno, conceptual knowledge and technical appropriation of the world rely on a quantitative and abstract levelling of individual natural qualities; the domination realises the uniformisation and homogenisation of a natural domain deprived of differences. The totalitarian character of the *Aufklärung* thus becomes evident in the modern era and finds its specific crystallisation – not traceable to the Pythagorean or Platonic view of mathematics as the key to understanding reality – in the conception of a homogeneous nature, determinable in the exact mode of objectivity thanks to mathematics, and in the expulsion of difference from this universe.[85] This homologation is based on the separation of the intellect from sensory experience in order to subjugate it. In turn, this separation leads to an impoverishment of both thought and experience. Thought that has lost the ability to reflect on itself is thought that is devoted solely to organisational and administrative tasks.[86] The elimination of qualities and the levelling of differences brought about by modern science result in the rationalisation of working methods, leading to an atrophy of people's perceptive functions. The regression of the masses is expressed in the inability of human beings, uniformed in a coercively directed community, to perceive anything new through their senses and to be no longer able to grasp the truth: 'The rowers, unable to speak to one another, are all harnessed to the same rhythms, like modern workers in factories, cinemas, and the collective'.[87] Machines bring about a paradoxical reconciliation between thought, crystallised in the material and intellectual apparatus, and living

85 'Thought is reified as an autonomous, automatic process, aping the machine it has itself produced, so that it can finally be replaced by the machine. Enlightenment pushed aside the classical demand to 'think thinking' – Fichte's philosophy is its radical fulfilment – because it distracted philosophers from the command to control praxis, which Fichte himself had wanted to enforce. Mathematical procedure [*Die mathematische Verfahrungsweise*] became a kind of ritual of thought [*zum Ritual des Gedankens*]. Despite its axiomatic self-limitation, it installed itself as necessary and objective: mathematics made thought into a thing – a tool, to use its own term'. Horkheimer and Adorno, *Dialektik der Aufklärung*, p. 48; *Dialectic of Enlightenment*, p. 19.

86 *Dialektik der Aufklärung*, p. 59; *Dialectic of Enlightenment*, p. 28.

87 *Dialektik der Aufklärung*, pp. 59–60; *Dialectic of Enlightenment*, pp. 28–9.

beings freed from certain exhausting tasks and duties; but in this reconciliation, the thought of an alienated reason becomes the uniform thought of society as a whole, not the individualised and divergent thought of its individual members. In other words, as Horkheimer and Adorno write in illustrating the epochal condition of alienated reason, 'the power of the system over human beings increases with every step they take away from the power of nature'; and the absurdity of this state of affairs 'denounces the reason of the reasonable society as obsolete'.[88]

The analysis now focuses on a long passage, previously quoted in part, in which Horkheimer and Adorno reveal the hidden connections between the definition of human beings in contrast to animals as the cornerstone of Western anthropology[89] and the experiments carried out by behaviourists on animals chained up in special laboratories.

> Throughout European history the idea of the human being has been expressed in contradistinction to the animal. The latter's lack of reason is the proof of human dignity. So insistently and unanimously has this antithesis been recited by all the earliest precursors of bourgeois thought, the ancient Jews, the Stoics, and the Early Fathers, and then through the Middle Ages to modern times, that few other ideas are so fundamental to Western anthropology. The antithesis is acknowledged even today. The behaviourists only appear to have forgotten it. That

88 *Dialektik der Aufklärung*, p. 62; *Dialectic of Enlightenment*, pp. 30–1.

89 In more recent times, Agamben – without referring to the Frankfurt School thinkers but placing his reflection in a broader framework extending from Pico della Mirandola to Heidegger – has developed the concept of the 'anthropological machine' that produces man through the opposition between man and animal in an amphibious relationship of inclusion/exclusion. 'Insofar as the production of man through the opposition man/animal, human/inhuman, is at stake here, the machine necessarily functions by means of an exclusion (which is also always already a capturing) and an inclusion (which is also always already an exclusion). Indeed, precisely because the human is already presupposed every time, the machine actually produces a kind of state of exception, a zone of indeterminacy in which the outside is nothing but the exclusion of an inside and the inside is in turn only the inclusion of an outside'. Giorgio Agamben, *The Open: Man and Animal*, trans. by Kevin Attell (Stanford, California: Stanford University Press, 2004), p. 37.

> they apply to human beings the same formulae and results which they wring without restraint from defenceless animals in their abominable physiological laboratories, proclaims the difference in an especially subtle way. The conclusion they draw from the mutilated animal bodies [*aus den verstümmelten Tierleibern*] applies, not to animals in freedom, but to human beings today. By mistreating animals, they announce that they, and only they in the whole of creation, function voluntarily in the same mechanical, blind, automatic way as the twitching movements of the bound victims made use of by the expert.[90]

In this paragraph Adorno and Horkheimer also focus on the oppression of animals and its specific role in the psychic mutilation of the human subject: in this light, vivisectionists are seen 'as the epitome of the distorted human being produced by enlightenment and its repressive mechanism'.[91] By mutilating the bodies of defenceless animals in the name of supposed rationality, those who conduct animal experiments are implementing a brutal and irrational logic that reduces all beings, human and non-human, to mere objects of calculation. The human control over internal and external nature in general and other animals in particular, which appears to be a triumph, has only come at the expense of further alienation and idiosyncrasy of human beings.[92] Indeed, Horkheimer and Adorno's strategy in approaching the core question they examine in their work involves uncovering a contradictoriness immanent in reason. Therefore, by proving that nature-dominating reason runs counter to its own purpose, they emphasize its irrational side: in other words, they highlight that reason is not a static, monolithic and one-sided faculty, but a multifaced strategy devised by human beings to face the dangers they perceive in the environment and control their fears. The legitimate tendency of conserving themselves regresses into the hubris of dominating nature which mutilates nature and human beings themselves. The helpless suffering of animals subjected

90 Horkheimer and Adorno, *Dialektik der Aufklärung*, p. 277–8; *Dialectic of Enlightenment*, pp. 204–5.

91 Zipporah Weisberg, 'Animal Repression: Speciesism as Pathology', in *Critical Theory and Animal Liberation*, ed. by John Sanbonmatsu (Plymouth: Rowman & Littlefield, 2011), pp. 177–193 (p. 179).

92 Weisberg, 'Animal Repression: Speciesism as Pathology', p. 179.

to vivisection, unable to escape their tormentors,[93] mirrors the submission of human beings to an irrational order. It is irrational because it ultimately conflicts with their natural tendency towards self-preservation and, moreover, it creates an alienated world in which reason fails to achieve its ends, namely universality (the happiness of all individuals, of society as a whole) and autonomy (the capacity for self-determination that makes society free).[94]

In a letter to Neal R. Healy dated 22 March 1945, Horkheimer explains that vivisection of animals is a clear sign of humans' inability to make proper use of the technical means necessary for their self-preservation. The laboratory where vivisection is practiced heralds the concentration camp: if nature becomes a mere object for human purposes, then some human beings too – 'denounced as members of a different race' – are considered objects to be banished.[95] 'Therefore – Horkheimer concludes – the fight against vivisection, the fight for the animal, is a fight for man and the bill before you has a greater civilizing function and meaning than many of its supporters may realize at this moment'.[96]

Animal testing is a telling example that tangibly and literally illustrates the theme of mutilation – of bodies and minds – that runs through the texts of the two authors. This concept recurs several times in their texts, both literally and meaningfully, and abstractly and metaphorically, to designate the suppression and control of the natural element in the social world. Consider for example a passage from 'Sketches and drafts on *Dialectic of Enlightenment*

93 'The professor at the dissection table defines such movements scientifically as reflexes; the soothsayer at the altar would have proclaimed them a sign from his gods. Humans possess reason, which pitilessly follows its path; the animals from which they draw their bloody conclusions have only unreasoning terror, the impulse to take flight on a path which is cut off'. Horkheimer and Adorno, *Dialektik der Aufklärung*, p. 278; *Dialectic of Enlightenment*, pp. 204.

94 On this point, see Horkheimer, *Nachtrag*, p. 221; 'Postscript', p. 248.

95 Horkheimer, *GS* 17: *Briefwechsel 1941-1948*, ed. by Gunzelin Schmid Noerr (Frankfurt am Main: Fischer, 1996), pp. 627–8; *A Life in Letters: Selected Correspondence*, ed. and trans. Manfred R. Jacobson and Evelyn M. Jacobson (Lincoln, NE: University of Nebraska Press, 2007), p. 227.

96 Horkheimer, *GS* 17, p. 628; *A Life in Letters*, p. 228.

[Aufzeichenungen und Entwürfe zur *Dialektik der Aufklärung*]' (1939–1942). Here Adorno and Horkheimer oppose theories that reinscribe everything in nature and reject the difference between nature and society: this difference emerges in the distinction between first and second nature. The first nature is what is generally and metaphorically mutilated in society, that is, in social nature:

> Peculiar requirements and their corresponding skills are formed and developed by individuals and groups in contrast to the rest of society, depending on the positions they hold in the new division of labour. Through their abilities, the occupation of key positions in the social apparatus, the hierarchy of power, originally based purely on natural potencies, has been modified to correspond, in the course of development, to second nature, that is, social nature: no longer a hierarchy of strength but of position. The first nature [*Die erste Natur*], on which society is still based, is mutilated and reduced to slavery [*wird verstümmelt und versklavt*].[97]

This passage explains that the social world is both social and natural, but nature appears as what is mutilated and subjugated in the process of construction of the society. The final outcome is the social world governed by exchange relations under the monopoly conditions characteristic of late capitalism, that appears to be natural.[98] Behind this fallacy, one can say that the first nature remains mutilated and reduced to slavery. 'First' nature is the material environment that gives rise to social relations ('second' nature): it 'must prefigure second in character. Thus, individuals come to find the social relations that constrict them already anticipated in the perceptible features of nature'.[99]

This, however, is not the only issue. The intertwining of the mutilation of nature and the mutilation of human beings manifests itself with particular emphasis in the subordinate position to which

97 Max Horkheimer, 'Aufzeichenungen und Entwürfe zur *Dialektik der Aufklärung*' (1939–1942), in *GS* 12, pp. 252–95 (p. 287). Translation by the author.

98 Cook, *Adorno on Nature*, p. 15.

99 Alison Stone, 'Adorno and the Disenchantment of Nature', *Philosophy and Social Criticism*, 2006, 32.2: 231–53 (p. 238).

women have been condemned in patriarchal society: 'She became an embodiment of biological function, an image of nature, in the suppression of which this civilization's claim to glory lay'. The patriarchal society disregards and stigmatises the weakness of woman: in a society where the goal is to dominate nature, 'biological inferiority remains the ultimate stigma, the weakness imprinted by nature, the mark which invites violence'.[100] Conversely, Horkheimer and Adorno comment, '[t]he oppressed woman as Fury has outlived her time, continuing to display the grimace of mutilated nature [*die Fratze der verstümmelten Natur*] in an age when domination is moulding the well-trained bodies of both sexes, in whose uniformity the grimace has been effaced'.[101] Here they refer to the more recent developments in patriarchal society, marked by an obsession with sculpted bodies. In their epoch, paradoxically, the obsessive concern for human bodies does not spare animals (pets, for example, which are not destined for slaughter):

> Like the hunchback's ungainly leaps, the little dog's features still represent mutilated nature [*die verstümmelte Natur*], while mass industry and mass culture have learned to prepare the bodies of breeding bulls and humans according to scientific methods. The standardized masses are now so little aware of their own transformation, in which they have desperately collaborated, that they no longer require its symbolic display.[102]

In summary, Horkheimer and Adorno highlight the mutilation as the specific character under which all natural entities find their unity. Natural beings are not unified by themselves, but by the domination imposed upon them and, in particular, they share a destiny marked by mutilation. Mutilation is both literary (bodily mutilation or modification) and metaphoric, that is, psychological or cultural. The fate of mutilation unites humans, animals and, in general, the

100 Horkheimer and Adorno, *Dialektik der Aufklärung*, p. 280; *Dialectic of Enlightenment*, p. 206.
101 Horkheimer and Adorno, *Dialektik der Aufklärung*, p. 282; *Dialectic of Enlightenment*, p. 208.
102 Horkheimer and Adorno, *Dialektik der Aufklärung*, p. 283; *Dialectic of Enlightenment*, p. 208.

environment.[103] In the case of human (auto)mutilation, it lies in the *entfremdete Ratio*, the alienated reason, which involves the loss of freedom that human beings experience in an over-technicised and rationalised world.[104] Thought – in the epoch of logistics, the era of large-scale forecasting and performativity at work – does not imply reflection on itself. It is a mere thought finalized to use the most efficacious tools to realise specific ends: a thought which is able to adapt itself to the system of domination and does not criticise this structure. In this context, the machinery, produced by a combination of human intelligence, science and technology, makes humans feel alienated and estranged from their own work. Self-mutilation is expressed, concretely, both in the repression of one's inner nature, impulses and instincts, and of self-reflection of reason. This is what Adorno also emphasises in *Minima Moralia* when describing the 'normality' of bourgeois life: 'The libidinal achievements demanded of an individual behaving as healthy in body and mind, are such as can be performed only at the cost of the profoundest mutilation [*der tiefsten Verstümmelung*], of internalized castration in extroverts, beside which the old renunciation of identification with the father is the child's play as which it was first rehearsed. The regular guy, the popular girl, have to repress not only their desires and insights, but even the symptoms that in bourgeois times resulted from repression'.[105]

Challenging the triumphant narratives of Western reason and the linear development of civilisation epitomised by the European model, Horkheimer and Adorno pursue a genealogical explanation

103 In Africa, for example, the environment has been modified and adapted to the specific needs of the rulers (both of nature and of the colonised peoples). 'The tall giraffe and the wise elephant are *oddities* which can no longer provide amusement for wised-up schoolboys. In Africa, the last part of the earth which has vainly sought to protect their poor herds from civilization, they form traffic obstacles to bombers landing in the latest war. They are being eradicated entirely'. Horkheimer and Adorno, *Dialektik der Aufklärung*, pp. 283–4; *Dialectic of Enlightenment*, p. 209.

104 See Christopher Rocco, 'Between Modernity and Postmodernity: Reading *Dialectic of Enlightenment* against the Grain', *Political Theory*, 22.1 (1994), 71–97 (p. 76).

105 Adorno, *Minima moralia*, § 36, p. 65; *Minima moralia*, English trans., p. 58.

of Western rationalism that uncover the mutual implication of enlightenment and myth, reason and domination, at the origin of civilization itself. Therefore, the human dominion over the Earth, which occurs through the transparency and controllability of nature for reified reason, is a structural tendency of enlightenment from the beginning. This being the case, enlightenment continues to legitimise domination over nature and interhuman domination until it questions its destructive aspect and the irrational consequences of its destruction of all mythology and metaphysics.[106] Enlightenment's lack of self-reflection proved fatal: only the restoration of such internal criticism, the 'reflection on [its] regressive moment', could reverse 'the self-destruction of enlightenment'.[107]

In the light of this analysis, *Dialectic of Enlightenment*'s critical potential seems to be well-founded. Far from having a pessimistic and fruitless outcome, 'it offers insights for engaging the present and the contemporary environmental crisis'.[108] The authors undermine the ideals of 'stringency and totality',[109] the myth of unilinear and unilateral progress, which is endorsed by the great Western narratives (including the philosophies of Habermas and Honneth); they show that this ideology ignores the mutilations, suffering, and impoverishment experienced by women and men during the advancement of the enlightenment, and that any critique of it must seriously consider the issue of the relationship between humans and nature as domination. They point to the need of a necessary rectification of the anthropocentrism and humanism of Western culture, which are also involved in the theories of communicative reason, discourse ethics and recognition, because the latter at best consider the environment as a backdrop for human activity and a minor concern. Such a correction does not imply restoring the cult

106 On this point, see Eric S. Nelson, 'Revisiting the Dialectic of Environment: Nature as Ideology and Ethics in Adorno and the Frankfurt School', *Telos*, 155 (Summer 2011): 105-126 (p. 110).

107 Horkheimer and Adorno, *Dialektik der Aufklärung*, pp. 18–19; *Dialectic of Enlightenment*, p. xvi.

108 Nelson, 'Revisiting the Dialectic of Environment', p. 126.

109 Adorno, *Minima moralia* § 98, p. 172; *Minima moralia*, English trans., p. 151.

of irrational life and vital power and the struggle for existence[110] but requires grasping the connection between the (social) domination of nature and the domination of humans over other humans.[111] Without claiming to be able to resolve the dilemma between natural abandonment to instincts (as a source of happiness) and the frustration of instincts that makes social security possible (which was thematised by Freud and would later be reconceptualised by Eric Fromm, who considers such an opposition superficial),[112] Horkheimer and Adorno believe that it is essential for (Western) men to become aware of the domination they have established over nature on a planetary level. This is the starting point for overcoming the paradox of man's domination of nature, which nevertheless enslaves him to it.[113]

---

110 'By leaving consideration of the destructive side of progress to its enemies, thought in its headlong rush into pragmatism is forfeiting its sublating character, and therefore its relation to truth. In the mysterious willingness of the technologically educated masses to fall under the spell of any despotism, in its self-destructive affinity to nationalist paranoia, in all this uncomprehended senselessness the weakness of contemporary theoretical understanding is evident'. Horkheimer and Adorno, *Dialektik der Aufklärung*, p. 19; *Dialectic of Enlightenment*, p. xvi.

111 Horkheimer and Adorno, *Dialektik der Aufklärung*, p. 20; *Dialectic of Enlightenment*, p. xvii.

112 See Erich Fromm, *The Sane Society* (1955) (London–New York: Routledge, 2002), pp. 72–5.

113 'By modestly confessing itself to be power and thus being taken back into nature [*Durch die Bescheidung, in der dieser als Herrschaft sich bekennt und in Natur zurücknimmt*], mind rids itself of the very claim to mastery which had enslaved it to nature'. 'Through this remembrance of nature within the subject [*Eingedenken der Natur im Subjekt*], a remembrance which contains the unrecognized truth of all culture, enlightenment is opposed in principle to power, and even in the time of Vanini the call to hold back enlightenment was uttered less from fear of exact science than from hatred of licentious thought, which had escaped the spell of nature by confessing itself to be nature's own dread of itself'. Horkheimer and Adorno, *Dialektik der Aufklärung*, pp. 63 and 64, respectively; *Dialectic of Enlightenment*, pp. 31 and 32, respectively.

3. *Reason's disease and earthly dominion by humans in* Eclipse of Reason

The theme of man's domination over nature, intrinsically intertwined with the domination of humans over other humans, is developed with a specific emphasis in a book by Horkheimer published in 1947: *Eclipse of Reaso*n. As is well known, it is a text based on a series of lectures Horkheimer gave at Columbia University in 1944. Therefore, it was elaborated around the same time as *Dialectic of Enlightenment.* But, as Abromeit notes, in the book co-written with Adorno, the critique seems to be generalised and 'tendentially dehistoricized', to the extent that the pregnant character of the instrumental, bourgeois rationality – that is the domination of external and internal nature – can already be found in classical antiquity. In contrast, one of the specificities of *Eclipse of Reason*, consists in the fundamental distinction between subjective and objective reason: the former characteristic is seen as the dominant trait of modern capitalist societies, the latter of classical antiquity.[114]

It should be noted that the perspective of *Eclipse of Reason* seems to shift from the idea of nature that emerged in the second half of the 1930s, when – in 'Traditional and Critical Theory' (1937) – Horkheimer seems to embrace the idea of a duality between human beings and nature found in Marx, Engels and the Marxist tradition of the 1930s:[115]

> There will always be something that is extrinsic to man's intellectual and material activity, namely nature as the totality of as yet unmastered elements with which society must deal. But when situations which really depend on man alone, the relationships of men in their work, and the course of man's own history are also accounted part of 'nature', the resultant extrinsicality is not only not a suprahistorical eternal category (even pure nature in the sense described is not that), but it is a sign of

114 John Abromeit, *Horkheimer and the Foundations of the Frankfurt School* (New York: Cambridge University Press, 2011), p. 401.

115 Helen Denham, 'The Cunning of Unreason and Nature's Revolt: Max Horkheimer and William Leiss on the Domination of Nature', *Environment and History*, 3.2, *Ecological Visionaries/Ecologised Visions* (June, 1997), 149–75 (pp. 153–4).

contemptible weakness. To surrender to such weakness is nonhuman and irrational.[116]

However, it is not certain that the first part of this passage suggests that 'non-human nature was an appropriate arena for human mastery',[117] not least because such a conception would be in contrast to the reflections on the continuity of suffering between non-human living beings and human beings that emerge in *Dämmerung*, published three years before 'Traditional and Critical Theory'. Furthermore, in a subsequent passage of the same essay, already mentioned,[118] Horkheimer states that 'after a period of progress, [of] development of human powers [*der Entfaltung menschlicher Kräfte*], and emancipation of the individual, after an enormous extension of human control over nature [*nach einer ungeheuren Ausbreitung der menschlichen Macht über die Natur*], [the historically given commodity economy] finally hinders further development and drives humanity into a new barbarism'.[119] The control and exploitation of nature certainly do not, in themselves, promote and produce human well-being and civilisation. Furthermore, nature – Horkheimer writes in the same essay – can never be totally controlled.[120] Of course, in this passage, Horkheimer also speaks of a struggle against nature which, in the rational society he advocates ('a future condition of things'), must become conscious. Nevertheless, the struggle

---

116 Horkheimer, 'Traditionelle und kritische Theorie', p. 184; 'Traditional and Critical Theory', p. 210.

117 Denham, 'The Cunning of Unreason and Nature's Revolt, p. 153.

118 See Chapter I, § 3, n. 63 of this book.

119 Horkheimer, 'Traditionelle und kritische Theorie', p. 201; 'Traditional and Critical Theory', p. 227.

120 'But to the extent that the subject does not totally isolate himself, even as thinker, from the social struggles of which he is a part and to the extent that he does not think of knowledge and action as distinct concepts, necessity acquires another meaning for him. If he encounters necessity which is not mastered by man, it takes shape either as that realm of nature [*das Naturreich*] which despite the far-reaching conquests still to come will never wholly vanish, or as the weakness of the society of previous ages in carrying on the struggle with nature in a consciously and purposefully organized way'. Horkheimer, 'Traditionelle und kritische Theorie', p. 204; 'Traditional and Critical Theory', p. 230.

against nature – which goes hand in hand with the struggle against the weakness of society – is not equivalent to its domination and uncontrolled exploitation but evokes the dialectical and mediated (and even antagonistic) relationship that exists between natural entities and human beings.[121]

Here, it is worth recalling briefly the fundamental thesis put forward in *Dialectic of Enlightenment*. The enlightenment aims to 'disenchant' the world, to free it from magical powers and myth, but ends up proposing new enchantments, thus falling back into mythology. At the beginning of the chapter 'The Concept of Enlightenment', Horkheimer and Adorno write, alluding to a formulation of Max Weber:[122] 'Enlightenment's program was the disenchantment of the world [*Die Entzauberung der Welt*]. It wanted to dispel myths, to overthrow fantasy with knowledge. The Enlightenment seeks to free man from slavery and domination, but instead reproduces new forms of slavery and strengthens domination over man'.[123] Horkheimer and Adorno clearly identify what they believe to be the key to understanding this paradoxical nature of the enlightenment. As previously observed, it lies in the origin itself of

---

121 'Both elements in this concept of necessity – the power of nature and the weakness of society – are interconnected and are based on the experienced effort of man to emancipate himself from coercion by nature and from those forms of social life and of the juridical, political, and cultural orders which have become a straitjacket for him. The struggle on two fronts, against nature and against society's weakness, is part of the effective striving for a future condition of things in which whatever man wills is also necessary and in which the necessity of the object becomes the necessity of a rationally mastered event'. Horkheimer, 'Traditionelle und kritische Theorie', p. 204; 'Traditional and Critical Theory', p. 230.

122 See Max Weber, 'Wissenschaft als Beruf' (1919), in Weber, *Gesammelte Aufsätze zur Wissenschaftslehre* (Tübingen: J.C.B Mohr (Paul Siebeck), 1968), p. 594. According to Alfred Schmidt, Weber's reflections form the background to the fundamental historical-philosophical thesis of Horkheimer and Adorno's book. See Alfred Schmidt, 'Aufklärung und Mythos im Werk Max Horkheimers', in *Max Horkheimer heute: Werk und Wirkung*, ed. by Alfred Schmidt and Norbert Altwicker (Frankfurt am Main: Fischer Taschenbuch Verlag, 1986), pp. 180–243 (pp. 196–7).

123 Horkheimer and Adorno, *Dialektik der Aufklärung*, p. 25; *Dialectic of Enlightenment*, p. 1.

the enlightenment process, that is, the roots of human civilisation. This began with a very specific act: the detachment of human beings from nature, their differentiation and opposition thereto. In Max Horkheimer's contemporary work, *Eclipse of Reason*, perhaps the most concise and effective formulation capable of showing this primordial action and its consequences on the process of civilisation is to be found:

> If one were to speak of a disease [*Krankheit*] affecting reason [*welche die Vernunft befällt*], this disease should be understood not as having stricken reason at some historical moment, but as being inseparable from the nature of reason in civilization as we have known it so far. The disease of reason is that reason was born from man's urge to dominate nature, and the 'recovery' [*die 'Genesung'*] depends on insight into the nature of the original disease, not on a cure of the latest symptoms. The true critique of reason will necessarily uncover the deepest layers of civilization and explore its earliest history. From the time when reason became the instrument for domination of human and extrahuman nature by man, that is to say, from its very beginnings it has been frustrated in its own intention of discovering the truth. This is due to the very fact that it made nature a mere object, and that it failed to discover the trace of itself in such objectivization, in the concepts of matter and things not less than in those of gods and spirit.[124]

On the one hand, as Lucio Cortella states,[125] Horkheimer's words clearly express the central idea contained in *Dialectic of Enlightenment*. Reason did not become diseased at a certain point in its development, but rather brought to its extreme conclusion[126]

124 Horkheimer, *Eclipse of Reason*, p. 176; *Zur Kritik der instrumentellen Vernunft*, p. 176.

125 Lucio Cortella, *Per una dialettica nella finitezza*, p. 8.

126 It is also true that this disease afflicting reason emerges symptomatically in the era of extraordinary achievements of modern industry and of increasing complexity of society. See Max Horkheimer, 'Die Vernunft im Widerstreit mit sich selbst. *Einige Bemerkungen zur Aufklärung*' (1946), *GS* 12, pp. 106–18 (p. 106); 'Reason Against itself: Some Remarks on Enlightenment', in James Schmidt (ed.), *What is Enlightenment? Eighteenth-century Answers and Twentieth-century Questions* (Berkeley–Los Angeles–London: University of California Press, 1996), pp. 359–67 (p. 359).

a disease it had contracted from its very birth. In fact, reason only comes into being when man begins to control, manipulate and dominate nature. Before that act, there is no reason because there is no differentiation between man and nature. Prior to this, the human being is nature.[127] And when he is no longer nature, he has already begun to be its dominator. On the other hand, recovery from disease depends on understanding the nature of the original illness of reason.[128] Such an understanding, in turn, requires that reason recognises its anchoring and rooting in nature, its connection with the instinct of self-preservation, which entails a tendency to dominate nature.[129] This idea, already expressed in the previous passage quoted, becomes clearer in the following sentences: 'The subjugation of nature will revert to subjugation of man, and vice versa, as long as man does not understand his own reason and the basic process by which he has created and is maintaining the antagonism that is about to destroy him. Reason can be more than nature only through concretely realizing its "naturalness" [*ihre 'Natürlichkeit'*] – which consists in its trend to domination – the very trend that paradoxically alienates it from nature'.[130]

As Alex Demirović affirms, if thought arose through the process of domination over nature and humankind, then it should also include criticism of domination at all times. According to Horkheimer and

127 On this point, see Fromm, *The Sane Society*, p. 24: 'Man's evolution is based on the fact that he has lost his original home, nature – and that he can never return to it, can never become an animal again. There is only one way he can take: to emerge fully from his natural home, to find a new home – one which he creates, by making the world a human one and by becoming truly human himself'.

128 'Reason can realize its reasonableness [*ihre Vernünftigkeit*] only through reflecting on the disease of the world as produced and reproduced by man; in such self-critique [*Selbstkritik*], reason will at the same time remain faithful to itself, by preserving and applying for no ulterior motive the principle of truth that we owe to reason alone'. Horkheimer, *Eclipse of Reason*, chapter 'The Concept of Philosophy', p. 177; *Zur Kritik der instrumentellen Vernunft*, p. 177.

129 See Franck Fischbach, *Pour la Théorie critique*, p. 29.

130 Horkheimer, *Eclipse of Reason*, p. 177; *Zur Kritik der instrumentellen Vernunft*, p. 177.

Adorno, individuals may emancipate themselves from the magical hold of nature on condition that thought or reason, through a process of self-reflection, comprehends itself as nature and at the same time reconnects, in some way, with a return to nature. This is clearly a specific form of self-reflection. The *Aufklärung*, realising that it is itself nature, discovers that science, technology, the ego, the subject born of the domination of natural impulses striving to preserve itself, are forms of the domination of nature. This domination project extends to the domination of human beings, the structuring of interpersonal relationships and the shaping of the way each individual thinks and behaves. It is only in this way, through its own self-criticism, that reason can ensure that its distance from nature and its ambition to dominate have always been historically false, and are no more relevant today.[131]

In *Eclipse of Reason*, Horkheimer provides a specific solution to save reason and the *Aufklärung* from the possibility of self-destruction. It consists in redirecting reason's tendency toward criticism in another direction, so that reason's self-destruction is countered by its self-critique. This is achieved to the extent that reason recognises the one-sidedness and limitations of the two opposing concepts of reason (subjective and objective): it is a matter of understanding their separation and their mutual relationship.[132] In particular 'the systems of objective reason' can correct the excesses of subjective reason, which, when taken to extremes, proves incapable of pursuing the goal it has itself highlighted: self-preservation. 'These metaphysical systems – Horkheimer writes – express in partly mythological form the insight that self-preservation can be achieved only in a supra-individual order, that is to say, through social solidarity'.[133] In other words, it is a

131 Alex Demirović, 'Raison et émancipation', in *La dialectique de la raison. Sous bénéfice d'inventaire*, ed. by Katia Genel (Paris: Éditions de la Maison des Sciences de l'Homme, 2017), pp. 87–104 (pp. 102–3).

132 Horkheimer, *Eclipse of Reason*, pp. 174–7; *Zur Kritik der instrumentellen Vernunft*, pp. 175–7. On subjective and objective reason, see the last paragraph of the second chapter of this book.

133 Horkheimer, *Eclipse of Reason*, p. 176; *Zur Kritik der instrumentellen Vernunft*, p. 176.

matter of correcting the definition of (bourgeois) reason in terms of individual self-preservation and overcoming its growing formal universality, which does not express 'an increasing consciousness of universal solidarity', but rather 'the sceptical separation of thinking from its object'.[134] However, according to Katia Genel, the knots do not seem to be easily untangled: the illness (and madness[135]) evoked by Horkheimer and Adorno[136] intimately concerns reason by virtue of its very functioning.[137] In fact, the solution consists in bringing the condition of nature – the non-human natural beings and the naturalness of human beings themselves – out of oblivion and the obvious. Compared to 'The End of Reason', *Eclipse of Reason* sheds light on a different managing of self-preservation in a broader framework of potential and uses of reason. The idea that reason destroys itself and turns into barbarism, in that it has always tended to reduce itself to an organ of self-preservation, whose original core – as Petrucciani rightly points out[138] – can be found in 'The End of Reason', should, however, be integrated and partially rectified with a view to saving the *Aufklärung*, in line with the perspective that

134 See Horkheimer, 'Vernunft und Selbsterhaltung', pp. 326–7; 'The End of Reason', p. 370.

135 'One might say that the collective madness [*der kollektive Wahnsinn*] that ranges today, from the concentration camps to the seemingly most harmless mass-culture reactions, was already present in germ in primitive objectivization, in the first man's calculating contemplation of the world as a prey.' Horkheimer, *Eclipse of Reason*, pp. 176–7; *Zur Kritik der instrumentellen Vernunft*, p. 176.

136 See for example Horkheimer and Adorno, *Dialektik der Aufklärung*, p. 230; *Dialectic of Enlightenment*, p. 165: 'By conquering the sickness of the mind [*der Krankheit der Geistes*] which flourishes on the rich soil of self-assertion unhampered by reflection, humanity would cease to be the universal antirace and become the species which, as nature, is more than mere nature, in that it is aware of its own image'.

137 Katia Genel, 'Une maladie de la raison? La métaphore médicale dans la théorie critique de Horkheimer', in *Genealogie della modernità: prospettive critiche*, ed. by Raffaele Carbone and Luca Scafoglio (Naples: FedOA Press, 2024), pp. 77–91 (pp. 90–1).

138 Stefano Petrucciani, *La Scuola di Francoforte* (Rome: Carocci, 2023), p. 45.

emerges with the diagnosis and prognosis formulated in *Eclipse of Reason*.

In this regard, it is important to examine the four 1946 discussion protocols entitled 'The Rescue of the Enlightenment [*Rettung der Aufklärung*]': here Horkheimer and Adorno again refer to the illness (*die Erkrankung*) of reason.[139] Adorno, in particular, putting the emphasis on the historical-social reality of transcendental-logical categories, intends to show 'the path by which one may cure reason of the illness [*von der Krankheit*] of its abstract absolutization'.[140] Indeed, in this October 1946 discussion protocol, Horkheimer and Adorno revisit the concept of the enlightenment, highlighting its positive aspect, and attempt to clarify their own position by showing that, far from being irrational, their project is about saving this concept. Here, Adorno states that Schopenhauer's central claim regarding the nothingness of the world 'is a conceptual repetition of the violence which human beings inflict upon nature'. And Horkheimer adds: 'It is the expression of violated nature. Schopenhauer's philosophy is misunderstood as ontology; it has been made out to be an ontology'.[141] In this protocol, Adorno offers an opposing view to Schopenhauer and Hegel, criticising 'the moment of identity philosophy'. According to him, Schopenhauer's nihilism reveals itself as an attempt to rescue 'the inner meaning of totality by declaring there is none'. But thought 'does violence to itself by saying "it's all for naught"'.[142] Consequently, rescuing *Aufklärung* implies rescuing difference. Indeed, Adorno distinguishes between a bad *Aufklärung* which, by virtue of its absolute totality, renounces the concept of difference, and a good *Aufklärung* that preserves difference in the face of levelling.[143] This means considering nature as the different, the non-identical, that cannot be reduced to identifying thought.

---

139 Horkheimer and Adorno, ['Rettung der Aufklärung. Diskussionen über eine geplante Schrift zur Dialektik'], *GS* 12, pp. 594–605 (pp. 602 and 604).

140 Horkheimer and Adorno, ['Rettung der Aufklärung'], pp. 603–4. For the English translation, see: https://jamescrane.substack.com/p/horkheimer-and-adornorettung-der.

141 Horkheimer and Adorno, ['Rettung der Aufklärung'], p. 596.

142 Horkheimer and Adorno, ['Rettung der Aufklärung'], p. 595.

143 Horkheimer and Adorno, ['Rettung der Aufklärung'], p. 595.

The texts examined – taken as a whole – seem to indicate that the disease of reason, even if it is original and consubstantial with reason, can find its cure. This therapy consists, of course, in the self-critique of reason (which serves as an antidote to the risk of falling into radical scepticism, and therefore nihilism), but this constant self-criticism, this vigilant attitude towards the risks associated with the functioning of reason, goes hand in hand with the development of an awareness of the nature that is denied and violated everywhere, and above all in human beings themselves. The thematization of the antagonism between human beings as rational beings and nature as that which is other than reason and denied by it, is a fundamental step in the therapy of reason. As Horkheimer writes in *Eclipse of Reason*, 'The possibility of a self-critique of reason presupposes, first, that the antagonism of reason and nature is in an acute and catastrophic phase, and, second, that at this stage of complete alienation the idea of truth is still accessible'.[144]

Remedying the oblivion of reason's original inscription in nature means grasping the connection between the separation of humans from nature and the fact that reason becomes purely instrumental reason and domination of nature; it also means shedding light on the paradox of a subjugation that ultimately reveals itself as powerlessness:

> Paradoxically, however, this increase of independence has led to a parallel increase of passivity. [...] Economic and social forces take on the character of blind natural powers that man, in order to preserve himself, must dominate by adjusting himself to them. As the end result of the process, we have on the one hand the self, the abstract ego emptied of all substance [*das abstract Ich, jeder Substanz entleer*] except its attempt to transform everything in heaven and on earth into means for its preservation, and on the other hand an empty nature degraded to mere material [*eine leere zu bloßen Material degradierte Natur*], mere stuff to be dominated, without any other purpose than that of this very domination.[145]

---

144 Horkheimer, *Eclipse of Reason*, p. 177; *Zur Kritik der instrumentellen Vernunft*, p. 177.

145 Horkheimer, *Eclipse of Reason*, p. 97; *Zur Kritik der instrumentellen Vernunft*, p. 109.

The paradox produced by the separation of humans from nature consists in a process of abstraction and emptying of both terms of the polarity. Human beings who adapt themselves to economic-social forces reveal themselves, in the end, to be an abstract emptied ego: they coincide with the mere strength for self-preservation, which allows them to use all natural resources in order to survive. Conversely, nature is also emptied and abstract, for it is reduced to a mere resource to be exploited. Not surprisingly, as Horkheimer and Adorno explain in *Dialectic of Enlightenment*, this separation leads to an ambivalent attitude – love/hate, attraction/repulsion – towards the body, which is seen and rejected as inferior and deserving of subjugation. 'In humanity's self-abasement to the *corpus* nature takes its revenge because the human being debases it to an object of power, to raw material'.[146]

This section will now consider in greater depth the third lecture of *Eclipse of Reason*, entitled 'The Revolt of Nature'. In this chapter Horkheimer highlights that, on the one hand, the domination of human beings over nature in modern times has produced the subjugation of nature in the individuals themselves (the internalisation of domination); on the other hand, it has deprived nature of its language.

> The human being, in the process of his emancipation, shares the fate of the rest of his world. Domination of nature involves domination of man. Each subject not only has to take part in the subjugation of external nature, human and nonhuman, but in order to do so must subjugate nature in himself. Domination becomes 'internalized' for domination's sake. What is usually indicated as a goal – the happiness of the individual, health, and wealth – gains its significance exclusively from its functional potentiality.[147]
>
> Once it was the endeavor [*das Bestreben*] of art, literature, and philosophy to express the meaning of things and of life, to be the voice of all that is dumb, to endow nature with an organ for making known her sufferings, or, we might say, to call reality by its rightful name. Today nature's tongue is taken away [*Heute ist der Natur die Sprache*

146 Horkheimer and Adorno, *Dialektik der Aufklärung*, pp. 264–5; *Dialectic of Enlightenment*, p. 193 (translation amended).

147 Horkheimer, *Eclipse of Reason*, p. 93; *Zur Kritik der instrumentellen Vernunft*, p. 106.

> *genommen*]. Once it was thought that each utterance, word, cry, or gesture had an intrinsic meaning; today it is merely an occurrence.[148]

Horkheimer evokes an epoch in which culture, endowed with critical spirit, was able to explore the secrets of nature without reducing it to mere, inert matter. However, in his time nature was deprived of its language. Natural beings are only an obstacle for the realisation of human projects or just a means to achieve a specific goal. 'The fate of animals in our world is symbolized by an item printed in newspapers of a few years ago. It reported that landings of planes in Africa were often hampered by herds of elephants and other beasts. Animals are here considered simply as obstructors of traffic. This mentality of man as the master can be traced back to the first chapters of Genesis'.[149] In the same text, Horkheimer later states:

> Nevertheless, nature is today more than ever conceived as a mere tool of man. It is the object of total exploitation that has no aim set by reason, and therefore no limit. Man's boundless imperialism is never satisfied. The dominion of the human race [*Die Herrshaft der menschlichen Gattung*] over the earth [*über die Erde*] has no parallel in those epochs of natural history [*der Naturgeschichte*] in which other animal species represented the highest forms of organic development. Their appetites were limited by the necessities of their physical existence. Indeed, man's avidity to extend his power in two infinities, the microcosm and the universe, does not arise directly from his own nature, but from the structure of society.[150]

Two fundamental considerations emerge from this passage. First, addressing nature as an object of human domination means developing full consciousness of how this domination came about, its extent and the absence of limits to human passions, particularly

148 Horkheimer, *Eclipse of Reason*, p. 101; *Zur Kritik der instrumentellen Vernunft*, pp. 112–13.

149 Horkheimer, *Eclipse of Reason*, p. 104; *Zur Kritik der instrumentellen Vernunft*, p. 115.

150 Horkheimer, *Eclipse of Reason*, p. 108; *Zur Kritik der instrumentellen Vernunft*, p. 119.

greed. Second, at this point an interesting element emerges: the boundless craving of human beings is not a natural fact, a direct consequence of their nature, but a historical product of a specific configuration and a particular development of society. This leaves open the possibility of a reconciliation between reason and nature, as Horkheimer clearly states in the same chapter: '[w]ithout committing the fallacy of equating nature and reason, mankind must try to reconcile the two'.[151] Conversely, as seen above, and in the previous chapter (§ 2), Horkheimer's understanding of the relationship between nature and reason is linked to critique, conceived as the self-critique of reason.[152]

Nevertheless, later, in the chapter 'The Concept of Philosophy', as discussed earlier, Horkheimer writes: 'Reason can be more than nature only through concretely realizing its "naturalness" [*ihre "Naturürlichkeit"*] – which consists in its trend to domination – the very trend that paradoxically alienates it from nature'.[153] Here Horkheimer seems to disclose the paradox of reason: human reason is also 'nature' but its 'naturalness' coincides with its tendency to dominate other beings and natural resources, which ends up opening a gulf between man and nature. Yet, in the chapter 'The Revolt

---

151 Horkheimer, *Eclipse of Reason*, p. 126; *Zur Kritik der instrumentellen Vernunft*, p. 134. A few lines before, Horkheimer writes: 'Instrumentalized subjective reason either eulogizes nature as pure vitality or disparages it as brute force, instead of treating it as a text to be interpreted by philosophy that, if rightly read, will unfold a tale of infinite suffering'. Horkheimer, *Eclipse of Reason*, p. 126; *Zur Kritik der instrumentellen Vernunft*, p. 134. Indeed, 'the only way for subjective reason to recover a sense of the objective was to have a memory of objectivity, of suffering, and to interpret it as a historical text'. Eric Oberle, *Theodor Adorno and the Century of Negative Identity* (Stanford, CA: Stanford University Press, 2018), p. 270.

152 On this point, see Katia Genel, *Autorité et émancipation. Horkheimer et la Théorie critique* (Paris: Payot, 2013), p. 162. Here, Genel observes that the path taken by Adorno is somewhat different: with him, the reconciliation between reason and nature can be understood as the establishment of a relationship with the other, by experiencing its fundamentally negative and mediated nature. On this point, see the next chapter of this book.

153 Horkheimer, *Eclipse of Reason*, p. 177; *Zur Kritik der instrumentellen Vernunft*, p. 177.

of Nature', as has been shown, in an almost Rousseauian vein, Horkheimer stated that human greed, man's avidity to dominate the microcosm and the universe, does not originate from his own nature, but from the structure of society – that is, from the form that human societies have historically taken. The problem lies, perhaps, in the fact that Horkheimer himself, despite repeatedly referring to 'nature', finds it difficult to define what 'nature' is. Of course, in 'The Concept of Philosophy', he states that neither can spirit be reduced to nature (and vice versa) nor can their polarity be hypostatised: 'spirit' and 'nature' are abstractions that imply 'a misrepresentation of concrete existence that ultimately affects the abstraction itself'. As a result, it is inadmissible to suppose an ultimate duality. 'The two poles cannot be reduced to a monistic principle [*ein monistisches Prinzip*], yet their duality [*Dualität*] too must be largely understood as a product'.[154]

In this regard, it may be interesting to note that, in the 1939 discussion protocol, Horkheimer and Adorno seem to agree that what separates the individual from nature is language. In particular, according to Horkheimer, the creation of myths seizes precisely upon extra-human nature, and even inorganic nature (for example, the sunset), in an undeniably animistic and personifying manner. Nevertheless, the creation of myth lies in language and thinking: 'Man – Horkheimer states – begins where language begins';[155] 'Man, through thinking, escapes the blind complex of nature [*dem blinden Naturzusammenhang*]'.[156] In his turn, Adorno emphasises the link between the myth and that which lacks language (nature): 'Myth (which means "word") is an exemplary situation because in myth, human beings try to account for the immediate experience

154 Horkheimer, *Eclipse of Reason*, p. 172; *Zur Kritik der instrumentellen Vernunft*, p. 171.

155 Max Horkheimer and Theodor W. Adorno, 'Diskussionen über die Differenz zwischen Positivismus und materialistische Dialektik', in Horkheimer, *GS* 12, pp. 437–92 (January 25, 1939, p. 460).

156 Horkheimer and Adorno, 'Diskussionen über die Differenz zwischen Positivismus und materialistische Dialektik', p. 461.

of what is devoid of language by means of language'.[157] As stated by Horkheimer, the myth explains that humans were once animals and became something else, while Adorno draws attention to the ambivalent relationship humans have with nature, a theme which, as we have seen, would be developed in *Dialectic of Enlightenment*: 'The myth attests both to the extent to which man inevitably succumbs to nature [*Naturverfallenheit*] and to his ability to free himself from it'.[158] Indeed, the 'word', the 'name' that takes shape in myths, expresses and condenses the original reaction to a natural reality perceived as threatening, unknown, unfamiliar.[159] This also gives rise to the belief that humans can free themselves from the strangeness and unfamiliarity of natural realities by understanding them and bringing them into the human realm of language and thought. This attitude has not only produced myth, but also the process of demythologisation, which is the very dynamic core of the *Aufklärung*: in this way, the enlightenment – which radicalises mythical fear – 'equates the living with the nonliving

157 Horkheimer and Adorno, 'Diskussionen über die Differenz zwischen Positivismus und materialistische Dialektik', pp. 460–1.

158 Horkheimer and Adorno, 'Diskussionen über die Differenz zwischen Positivismus und materialistische Dialektik', p. 461.

159 'The cry of terror called forth by the unfamiliar [*das Ungewohnte*] becomes its name. It fixes the transcendence of the unknown in relation to the known, permanently linking horror to holiness. The doubling of nature into appearance and essence, effect and force, made possible by myth no less than by science, springs from human fear, the expression of which becomes its explanation'. Horkheimer and Adorno, *Dialektik der Aufklärung*, p. 37; *Dialectic of Enlightenment*, p. 10. Nevertheless, as Adorno states in another 1939 discussion protocol, through naming, objects lose the frightening character [*Schrecken*] they have as objects of nature but acquire the frightening character of social objects. Max Horkheimer and Theodor W. Adorno, 'Diskussionen über Sprache und Erkenntnis, Naturbeherrschung am Menschen, politische Aspekte des Marxismus', in Horkheimer, *GS* 12, pp. 494–525 (p. 504). More recently, and with a different emphasis, Lovelock has highlighted the difficulty of putting the sphere of life and natural entities into words: 'Life, the universe, consciousness, and even simpler things like riding a bicycle, are inexplicable in words'. James Lovelock, *The Revenge of Gaia: Earth's Climate in Crisis and the Fate of Humanity*, foreword by Sir Crispin Tickell (London: Penguin Books, 2007 [2006]), p. 49.

as myth had equated the nonliving with the living'.[160] Questioning the original nexus between man and nature, with the awareness that this connection must be understood in dialectical terms (there is no original nature and then a subject that enters into a relationship with it), Horkheimer and Adorno highlight the difficulty of illuminating such a connection, as if the process of nature's self-clarification through man (as part of it) were still ongoing. 'I believe – Horkheimer states in the 1939 discussion protocol – that all of nature is waiting, so to speak, for us to find the right word. Every man is, at heart, a trial run, and it has taken an eternity to get here. A dark mass [*Eine dunkle Mass*]; from time to time, a few sparks glow, and then it must happen.'[161] However, it can be argued – as Horkheimer states in a 1946 text[162] – that fear of nature, conceived as an immeasurable and unpredictable Being, 'has been replaced by confidence in abstract formulae'. Indeed, over the centuries, nature has changed 'its appearance [*ihr Erscheinungsbild*]': terrible and overwhelming entity Mana (preanimistic age), nebulous and indefinable various spirits and gods (animistic age), gods with clear contours (epic poems), basic elements of all things (Pre-Socratic philosophers), eternal concepts and ideas (Plato), 'dead matter – a heap of things' (modern science). In this final metamorphosis of its appearance, nature lost every mark of 'vital independent existence', all intrinsic value.[163]

---

160 Horkheimer and Adorno, *Dialektik der Aufklärung*, p. 38; *Dialectic of Enlightenment*, p. 11.

161 Horkheimer and Adorno, 'Diskussionen über die Differenz zwischen Positivismus und materialistische Dialektik', p. 463.

162 The typescript is entitled 'The Reason Against Itself. Some Remarks on Enlightenment,' dated 1946, MHA: X 1a. We can deduce from a programme of the American Philosophical Association that Horkheimer gave his lecture on 28 December 1946 at the association's 20th Annual Meeting at the University of Oregon in Eugene. See Horkheimer, *GS* 12, 'Editorische Vorbemerkung', p. 105.

163 Horkheimer, 'Die Vernunft im Widerstreit mit sich selbst. *Einige Bemerkungen zur Aufklärung*', p. 109; 'Reason Against itself: Some Remarks on Enlightenment', p. 361. Note that, in the 1939 discussion protocol, concerning the mythical phase of humans-nature relationship, Horkheimer affirms: 'The real relationship between man and nature has historically been transformed into a relationship between the one who

A number of theoretical elements identified in the theses examined in this chapter lead us to propose, in conclusion, a possible comparison between the analysis of Horkheimer and Adorno and the theoretical perspective developed over the last two decades by Bruno Latour. The three authors converge in their criticism of the view of nature developed by modern thinking, which for Horkheimer and Adorno is the age dominated by subjective reason. According to Bruno Latour, the theological-epistemological conception of causality in modern philosophy and science, that is 'the causalist narrative', produces an 'effect of deanimation'; it deprives natural beings of causal power and agency.[164] Horkheimer/Adorno and Latour admit that the end of animism is a part of the disempowerment of nature, of the deprivation of its agency. Horkheimer and Adorno's issue seems to be recognising, through a genealogical analysis, that, in modern times, nature has become mute, inert matter, in relation to a subject that no longer knows how to interpret it and merely exploits it for its own ends. Conversely, for Latour, who echoes Lovelock's (and Lynn Margulis') 'Gaia hypothesis'[165] and highlights its misunderstandings

remembers and that which is remembered. At the stage of myth formation, man experiences nature in the form of memory'. Horkheimer and Adorno, 'Diskussionen über die Differenz zwischen Positivismus und materialistische Dialektik', p. 460.

164 Latour, *Face à Gaïa*, p. 94; *Facing Gaia*, p. 71.

165 Consider this definition of the concept of 'Gaia': 'We have to think of Gaia as the whole system of animate and inanimate parts. The burgeoning growth of living things enabled by sunlight empowers Gaia, but this wild chaotic power is bridled by constraints which shape the goal-seeking entity that regulates itself on life's behalf. I see the recognition of these constraints to growth as essential to the intuitive understanding of Gaia. Important to this understanding is that constraints affect not only the organisms or the biosphere but also the physical and chemical environment'. Lovelock, *The Revenge of Gaia*, p. 20. Lovelock uses the metaphor of 'the living Earth' to definite Gaia, but he does not assume he is thinking of the Earth 'as alive in a sentient way, or even alive like an animal or a bacterium'. See Ibid. According to Lovelock, 'Gaia' is just a metaphor – not to be taken literally (ibid.) –, but this metaphor is important because allow us 'to know the true nature of the Earth and imagine it as the largest living thing in the solar system, not something inanimate like that disreputable contraption "spaceship Earth"'. Ibid., p. 21.

by its critics, it is crucial to recognise that the Earth reacts to certain actions and behaves as a self-regulating system. The 'Gaia hypothesis' proposes that the Earth is a living being of which humans are an integral part. The planet Earth is not just one body among others, but completely unique. Far from being a collection of Galilean objects, the Earth is a world of agents or actors with whom humans are in exchange. Latour designates 'the Terrestrial as a new political actor' and peremptorily affirms: 'what is changing is that, henceforth, "geo" designates an agent that participates fully in public life'.[166]

Indeed, for Latour the core problem is the fact that the Galilean solution consisting in removing all behaviours from bodies to retain only their movement, crystalises in modern epistemology and metaphysics (Descartes, Locke) which does not attribute any agency or power to bodies, to Earth.[167] 'It is nevertheless this unwarranted generalization that gave rise to the strange opinion that has made it possible to deanimate one sector of the world, deemed objective and inert, and to overanimate another sector, deemed to be subjective, conscious, and free'.[168] In this regard, Latour introduces the concept of 'metamorphic zone'.

> The other hypothesis consists in proposing that what I have designated as a zone of common exchange – that is, the metamorphic zone – is *a property of the world itself* and not only a phenomenon of language *about* the world. Even if it is always difficult to keep this in mind, the analysis of meaning – the science of meaning, or semiotics – has never been limited to discourse, language, texts, or fictions. Signification is a property of all agents, in that they never cease to have agency; this is equally true of Kutuzov, the Mississippi, the CRF receptor, and the gravity through which bodies 'comprehend' and mutually 'influence' one another.[169]

---

166 See also Bruno Latour, *Down to Earth: Politics in the New Climatic Regime* (Cambridge: Polity, 2018), p. 40.

167 Latour, *Face à Gaïa*, p. 111; *Facing Gaia*, p. 85.

168 B. Latour, *Face à Gaïa*, pp. 111–12; *Facing Gaia*, p. 85.

169 B. Latour, *Face à Gaïa*, p. 92; *Facing Gaia*, p. 69.

It has been noted that, according to Adorno and Horkheimer, the vision and manipulation of nature in modern times, in the epoch of bourgeois society, ends up characterising it as blind, mute, mutilated. Like Latour, the two German thinkers believe that nature, manipulated by modern science and technology, is deprived of sensitivity and life. This relationship to nature lies on the Galilean schema, i.e. on the distinction between primary and secondary qualities, indiscriminately extended to the whole of reality. Galileo removed all behaviours from bodies and maintained only their movement, but his successors (in the first place, Descartes and Locke) turned this practice into a general philosophy and 'still less into the politics of an Earth deprived of any possibility of being moved'.[170] Indeed, as Husserl states in his famous book, which Horkheimer and Adorno drew on in order to describe the modern reduction of nature and natural entities to mere equivalent quantities in *Dialectic of Enlightenment*, 'with Galileo, then, begins the surreptitious substitution of idealized nature for prescientifically intuited nature'.[171] For Latour, the issue is not only about recovering the pre-scientific (pre-modern) experience of nature or denouncing the fact that, in the modern technical-scientific view, nature is mutilated, mute and blind (yet still capable of making its presence felt, precisely through the marginal and reductive condition to which it has been relegated, according to Horkheimer and Adorno). Rather, Latour maintains that the problem is to call into question the belief in a world pretentiously considered to be lifeless and highlighting animation as an essential phenomenon that characterises natural entities: notably, bringing out their potential for action, their agency.[172]

170 Latour, *Face à Gaïa*, p. 111; *Facing Gaia*, p. 85.

171 Husserl, *The Crisis of European Sciences and Transcendental Phenomenology*, pp. 49–50.

172 Latour, *Face à Gaïa*, pp. 93–4; *Facing Gaia*, pp. 70–1. For Latour, '[d]eanimation […] is not a primary process but, rather, a secondary treatment, polemical and apologetic, which attributes to the sciences and the world they describe the behavior of inert and obtuse things that are as unlike them as the overanimation proposed by their adversaries'. Latour, *Face à Gaïa*, p. 214; *Facing Gaia*, p. 170.

IV.
THE END OF BOURGEOIS HUMANISM AND THE MINDFULNESS OF NATURE IN ADORNO'S WRITINGS OF THE 1950S AND 1960S

1. *Humanism and nature in* Minima moralia

Before discussing Adorno's position on humanism and nature in more detail, particularly in his post-*Dialectic of Enlightenment* writings, it is useful to briefly review the attention that critical literature has devoted to these themes in Adorno, specifically that of nature. The last thirty years have seen a renewed interest in the conception of nature in Adorno's work, also in view of its possible use in ecological discourse. This interest was already evident in the 1990s in the works of Jameson and Vogel: while the former considers Adorno and the other members of the Institute for Social Research as precursors of the ecological movement and animal rights struggles, the latter believes that Adorno (and Horkheimer) fails to conceive of a dialectical interaction between nature and society, remaining tied to a vague and ambiguous notion of nature: Nature is regarded as a given reality that humans seek to dominate, but it is also what enlightenment would like to return to.[1] While, as shown earlier in this work, more recent scholars such as Cook, Nelson and Vuillerod[2] have attempted to define the contours and

1 Fredric Jameson, *Late Marxism: Adorno, or, the Persistence of the Dialectic* (London: Verso, 1990), p. 96; Vogel, *Against Nature*, pp. 62–3.

2 See also Sabine Wilke, 'Enlightenment, Dialectic, and the Anthropocene: Bruised Nature and the Residues of Freedom', *Telos*, 177 (2016), 83–106; Alastair Morgan, 'Reconciliation with Nature: Adorno on Reason, Nature and Critique', *Adorno Studies*, 3.1 (2019), 20–32; Agnès Grivaux, 'Natur zwischen Übermacht, Unschuld und Instrument, Zur Skizze eines minimalen Naturbegriffs bei Adorno', *Deutsche Zeitschrift für Philosophie*, 69.6 (2021), 989–1002.

importance of the concept of nature in Adorno, it should also be noted that Adorno himself does not provide a precise definition of the concept of nature, but uses it in a vague, sometimes elusive way, consistent with his distrust of philosophical thinking based on rigid conceptual definitions.[3] Of course, the concept of nature is not the object of a specific and detailed investigation in Adorno – here too, interpreters agree – but it is not scattered inconsistently throughout his work, particularly – as discussed in the previous chapter – if nature is considered the object of domination in bourgeois society. Therefore, beyond the problematic and ambiguous character of the concept of nature itself – and in view of the fact that for Adorno, as for Horkheimer, nature in itself, separated from the processes of social mediation, does not exist[4] – the notion of nature remains paradoxically central to Adorno, insofar as nature is what obsesses human beings and dominates them. Furthermore, the interaction between nature and domination always constitutes a litmus test

---

3 See Dominic Roulx, 'Le concept de nature chez Adorno', *Laval théologique et philosophique*, 79.3 (2023), 435–49 (pp. 437– 38).

4 On the mediate character of nature, see for example: Theodor W. Adorno, *Ästhetische Theorie*, ed. by Gretel Adorno and Rolf Tiedemann (Frankfurt am Main: Suhrkamp, 1973), p. 107; *Aesthetic Theory*, ed. by Gretel Adorno and Rolf Tiedemann, newly translated, edited, and with a translator's introduction by Robert Hullot-Kentor (London–New York: Continuum, 2002), p. 68: 'For in every particular aesthetic experience of nature the social whole is lodged. Society not only provides the schemata of perception but peremptorily determines what nature means through contrast and similarity. Experience of nature is coconstituted by the capacity of determinate negation. With the expansion of technique and, even more important, the total expansion of the exchange principle, natural beauty increasingly fulfills a contrasting function and is thus integrated into the reified world it opposes. Coined in opposition to absolutism's wigs and formal gardens, the concept of natural beauty forfeited its power, because bourgeois emancipation under the sign of the alleged natural rights of human beings made the world of experience not less but more reified than it was in the eighteenth century. The unmediated experience of nature, its critical edge blunted and subsumed to the exchange relation such as is represented in the phrase "tourist industry," became insignificantly neutral and apologetic, and nature became a nature reserve and an alibi. Natural beauty is ideology where it serves to disguise mediatedness as immediacy'.

that reveals the contradictions and decadence of anthropocentric humanism produced by bourgeois society.[5]

In some of his post-war writings, particularly in *Minima moralia*, Adorno accurately reconstructs the shattering of bourgeois ideals and the alienation of the individual – the self-determining subject of modern times, apparently heralded by Pico della Mirandola and embodied by Francis Bacon and Descartes on an epistemological-theoretical level – in the most subtle ramifications of everyday life.[6] Adorno demonstrates that such a presumed self-determining subject does not have a reliable integrity but is entirely governed by societal forces alien to it – unfamiliar even if they seem natural –, highlighting the link between the estrangement in habits and gestures of everyday life, and the frustrated desire to dominate and control nature that characterises modernity and capitalist society.[7] Aphorism 53 from *Minima Moralia* is illuminating in this regard. Here, Adorno draws attention to the intimate connivance between the tendency of the idealist to embrace the whole of humanity – while being inclined 'to trample inhumanely' on what is small as mere existence –, and the

5 According to Arnd Hoffmann, Adorno develops a representation of nature as a construction of immanent social relationships. In it, animals are conceived as socialised beings or within the context of a philosophy of nature that seeks to understand humans as animals. This dual approach should avoid the anthropocentric exclusivity of humans. Adorno's thinking is therefore oriented 'in an "anti-humanistic" way', because it frees both animals from the mechanistic attribution of pure naturalness and humans from the mistaken belief that they are separate from nature. In brief, society embraces nature just as nature itself dominates society. Arnd Hoffmann, 'Adorno und die Tiere', in *Adorno-Handbuch: Leben – Werk – Wirkung*, ed. by Richard Klein, Johann Kreuzer and Stefan Müller-Doohm, 2nd edition (Berlin: J. B. Metzler, 2019), pp. 444–53 (pp. 445–6).

6 'Whatever good suggestions the reader may find in *Minima Moralia*, it is framed by repeated, even obsessive, announcements of the end of the bourgeois era as well as of the demise of the self-determining subject, the projected recipient of advice'. Jakob Norberg, 'Adorno's Advice: *Minima Moralia* and the Critique of Liberalism', *Publications of the Modern Language Association* (*PMLA*), 126.2 (2011), 398–411 (p. 400).

7 See Rahel Jaeggi, '"No Individual Can Resist": *Minima Moralia* as Critique of Forms of Life', trans. by James Ingram, *Constellations*, 12.1 (2005), 65–82 (pp. 65–6).

vulgar need for ostentation of magnates and powerful bourgeois.[8] Adorno reflects on the way in which bourgeois culture seeks to exorcise its vulgarity and mediocrity by transfiguring its pretension to dominate nature into an idealised and stylised dimension. Yet classicist idealisation, which reveals the absence of an authentic experience of the transfigured natural entity, proves to be nothing more than a clumsy and unsuccessful attempt to hide the modest objects and ordinary sensations of everyday life:

> In the classical apotheosis of the sovereignty over nature, the vulgar and inferior mirrors itself by assiduous negation. Close behind the ideal stands life. The rose-scents of Elysium, much too voluble to be credited with the experience of a single rose, smell of the tobacco-smoke in a magistrate's office, and the soulful moon on the backdrop was fashioned after the miserable oil-lamp by whose meagre light the student swots for his exam. Weakness posing as strength betrayed the thought of the allegedly rising bourgeoisie to ideology, even when the class was thundering against tyranny. In the innermost recesses of humanism [*Im innersten Gehäuse des Humanismus*], as its very soul, there rages a frantic prisoner [*tobt gefangen der Wüterich*] who, as a Fascist, turns the world into a prison [*zum Gefängnis*]."[9]

Humanism and bourgeois ideals in general conceal a weakness that masquerades as strength – corresponding to the attempt to find beauty even in what is dull or indifferent, or to affirm the sacred nature of life in what is ugliest or most deformed.[10] This weakness, to the extent that it becomes consciousness of its daily frustration and alienation, painfully experiences the anguish of the prison built by bourgeois society, whose ultimate outcome is the frenzy and madness of fascism. The exaltation of life and all its manifestations – a typical tendency of fascist regimes and a reaction to the atomisation and empty rationalism of bourgeois society – contributes to depriving

8 Adorno, *Minima moralia.* § 53, p. 99; *Minima moralia*, English trans., p. 88.

9 Adorno, *Minima moralia.* § 53, pp. 99–100; *Minima moralia*, English trans., p. 89.

10 Adorno, *Minima moralia.* § 48, p. 86; *Minima moralia*, English trans., p. 77.

others (those who are different) of the possibility of light and the possibility of existing in the blind impulse of self-affirmation (*in blinder Selbstbehauptung*).[11] Even the abyss of the self, the cases and impulses investigated by psychoanalysis, should no longer surprise or frighten the bourgeois subject. They are reduced to common formulas and conventions, to examples of publicly recognised constellations; in this way, conflicts are normalised and accepted, but not cured.[12] Therefore, '[t]he last grandly-conceived theorem of bourgeois self-criticism has become a means of making bourgeois self-alienation [*die bürgerliche Selbstkritik*], in its final phase, absolute, and of rendering ineffectual the lingering awareness of the ancient wound, in which lies hope of a better future'.[13] As the dissection of human beings into their faculties is a projection of the division of labour between different subjects, aimed at more effective manipulation of their lives, similarly the dismantling of the unity of the self by psychoanalysis contributes to undermining the autonomy of individuals and, therefore, to subjecting them to the mechanism of rationalisation, to encouraging their adaptation to the existing system.[14] In this regard, Adorno believes that 'Hume, whose work bears witness in every sentence to his real humanism [*realen Humanismus*], yet who dismisses the self as a prejudice, expresses in this contradiction the nature of psychology as such'.[15]

The bourgeois (self)alienation evoked in aphorism 40 also includes a false notion of nature, which is in reality 'the scar of social mutilation [*das Wundmal gesellschaftlicher Verstümmelung*]'.[16] It reveals itself in the feminine character and the ideal of femininity which appeals to instinct, produced by masculine society and subjugated to male

11 Adorno, *Minima moralia.* § 48, p. 87; *Minima moralia*, English trans., p. 77.

12 Adorno, *Minima moralia.* § 40, pp. 72–3; *Minima moralia*, English trans., pp. 65–6.

13 Adorno, *Minima moralia.* § 40, p. 74; *Minima moralia*, English trans., p. 66.

14 Adorno, *Minima moralia*, § 39, p. 71; *Minima moralia*, English trans., pp. 63–4.

15 Adorno, *Minima moralia*, § 39, p. 71; *Minima moralia*, English trans., p. 64.

16 Adorno, *Minima moralia.* § 59, p. 107; *Minima moralia*, English trans., p. 95.

violence.[17] The lie is revealed, first of all, in the fact that, in bourgeois society, nature exists to the extent that it is framed within the system, and thus tolerated. Secondly, and more radically, in the fact that what appears as nature in civilisation is, in reality, the opposite of nature, that is, pure objectification, 'its own self-chosen object'.[18] In fact, as Adorno writes in the last aphorism of the second part – and on this point he seems to agree with Lukacs' position on the matter in *History and Class Consciousness*[19] –, the bourgeois concept of nature has always aimed to sanction social violence as unchangeable, 'as a piece of healthy eternity', to disguise the social element in 'universal nature'.[20] Consistent with this diagnosis, in the aforementioned and analysed aphorism (number 74),[21] Adorno highlights the paradox of a nature merely surviving in an irrational culture (which is integrated into the apparatus of domination and the conventions of 'second nature') and a rationalised culture that absorbs nature into itself (to the point of domesticating wild animals), thus eliminating nature (difference) and, with it, the possibility of reconciliation between

17 Adorno, *Minima moralia.* § 59, p. 107; *Minima moralia*, English trans., pp. 95–6.

18 Adorno, *Minima moralia.* § 59, p. 107; *Minima moralia*, English trans., pp. 95–6.

19 'For the individual, whether capitalist or proletarian, his environment, his social milieu (including Nature which is the theoretical reflection and projection of that milieu) must appear the servant of a brutal and senseless fate which is eternally alien to him. This world can only be understood by means of a theory which postulates "eternal laws of nature". Such a theory endows the world with a rationality alien to man and human action can neither penetrate nor influence the world if man takes up a purely contemplative and fatalistic stance'. Georg Lukács, *Geschichte und Klassenbewußtsein. Studien über marxistische Dialektik* (Bielefeld: Aisthesis Verlag, 2013), p. 211; *History and Class Consciousness: Studies in Marxist Dialectics*, trans. by Rodney Livingstone (Cambridge, MA: The MIT Press, 1971), p. 38. On the '"ewige" Naturgesetze', see also Lukács, *Geschichte und Klassenbewußtsein*, p. 219; *History and Class Consciousness* p. 47.

20 Adorno, *Minima moralia.* § 100, p. 178; *Minima moralia*, English trans., p. 156.

21 See chapter I, § 2 of our book.

nature and culture.[22] Conversely, in the culture framed and regulated by the cultural industry system, the ambivalent progress of bourgeois civilisation has constructed the palace of Janus (from aphorism 96), in which it cultivates both 'the potential of freedom' and 'the reality of oppression', thus giving rise to 'a situation where peoples were more and more inducted into the control of nature [*immer vollständiger der Naturbeherrschung*] and social organization' and making them incapable, under the coercion of culture, 'of understanding in what way culture went beyond such integration'.[23] In other words, paradoxically, '[w]hat has become alien [*fremd*] to men is the human component of culture [*das Menschliche an der Kultur*], its closest part, which upholds them against the world'.[24] The adjective 'fremd' refers to that which is unknown, the condition of being a stranger and that of alienation (referring to something that passes into the hands of others). The crisis of bourgeois humanism is expressed precisely in this triple estrangement, whereby great works of art and philosophical systems remain uncomprehended due to their excessive proximity[25] to a human experience alienated in the workings of a system which, by reducing nature to an object to be dominated, also controls and dominates human beings in their working and leisure time. Paradoxically, such an incomprehension could conceal a 'too great comprehension: shame at involvement in universal injustice that would become overwhelming as soon as one allowed oneself to understand'.[26] What is called 'culture' in the Western tradition – Adorno states in the concluding part of this aphorism – now administers the whole of humanity and, in this way, also administers the breach between man and culture, incorporating

22 Adorno, *Minima moralia*, § 74, pp. 131–2; *Minima moralia*, Engl. trans., p. 116.
23 Adorno, *Minima moralia*, § 96, p. 167; *Minima moralia*, Engl. trans., p. 146.
24 Adorno, *Minima moralia*, § 96, p. 167; *Minima moralia*, Engl. trans., p. 146.
25 'not through their too great distance from the heart of human experience, but the opposite'. Adorno, *Minima moralia*, § 96, p. 167; *Minima moralia*, Engl. trans., p. 147.
26 Adorno, *Minima moralia*, § 96, p. 167; *Minima moralia*, Engl. trans., p. 147.

barbarism into culture. The mass society created by decision-makers 'did not first produce the trash for the customers, but the customers themselves'.[27]

2. Humanität *and nature in 'Trying to Understand* Endgame*'*

Although *Minima Moralia* was written between 1944 and 1947, shortly after *Dialectic of Enlightenment*, and published in Germany in 1951, the themes addressed in this book – and its 'negativistic approach'[28] – were developed further in subsequent texts. This section focuses on an article from the early 1960s: 'Trying to Understand *Endgame* [Versuch das *Endspiel* zu verstehen]' (1961), which is part of Adorno's collection of essays *Noten zur Literatur*.[29] This text consists of three parts: *Noten zur Literatur I* appeared in 1958, *Noten zur Literatur II* in 1961, and *Noten zur Literatur* III in 1965.

It is useful to summarise briefly the contents of Beckett's play.[30] The action unfolds in a single room and involves two main figures: Hamm, an old blind man who cannot stand, and his servant Clov, who, conversely, is unable to sit. Bound to one another, they have spent many years quarrelling and continue this conflict throughout the play. Clov repeatedly expresses a desire to leave, yet appears incapable of doing so. Also present on stage are Hamm's very elderly parents, Nagg and Nell, both without legs and living in rubbish bins placed at the front left of the stage. The characters exist in a fixed, unchanging condition in which each day repeats the last, so that every action and occurrence acquires an almost ceremonial quality. The title *Endgame* comes from the final phase of a chess match, when only a few pieces remain on the board. This chess metaphor suggests

27 Adorno, *Minima moralia*, § 96, p. 168; *Minima moralia*, Engl. trans., pp. 147–8.

28 Jaeggi, '"No Individual Can Resist": *Minima Moralia* as Critique of Forms of Life', p. 75.

29 See Dirk van Hulle, 'Adorno's Notes on *Endgame*', *Journal of Beckett Studies*, 19.2 (2010), 196–217.

30 Samuel Beckett, *Fin de partie* (Paris: Les Éditions de Minuit, 1957).

a world in which everyone is a powerless pawn, waiting for the end in a kind of drowsy watchfulness broken only by brief memories. Beckett's play is a parody, a form uniquely suited to satirizing an age shaped by existentialism, with its focus on individuality, the isolation of the self in the face of the world, futility, instability, failure, the absurd nature of existence, and the boundaries and possibilities of individual freedom, all ultimately revolving around the question of what it means 'to exist'.

Adorno detects in *Endgame* a critique of existentialism, a statement of the destruction of the bourgeois subject (and any philosophy that attempts to revitalise it) and, finally and perhaps most importantly, an examination and indictment of post-Holocaust culture. Indeed, 'there is a literal but unspecified catastrophe that forms not only the background to the play but is the constant shadow of all the dialogue and of all the action'.[31] As scholars have often argued, the play could be taking place in a post-Nuclear world, though – as Matthew Holt points out – there is no direct evidence for this interpretation. 'But what is clear is that there has been devastation; of what kind and for what reasons we are not told'.[32]

The catastrophe that inspired Beckett's *Endgame* shattered this individual whose substantiality and absoluteness were the common thread linking Kierkegaard, Jaspers and Sartre's version of existentialism. *Endgame* destroys the illusion of freedom, substance and identity of the individual. This play refutes the individualistic position defended by the ontological enterprise of all existentialism, even that of *Being and Time*.[33] Adorno thus develops his interpretation of Beckett's text in relation to the major philosophical trends of the twentieth century:

> Beckett's drama abandons that position like an outmoded bunker. If individual experience in its narrowness and contingency has interpreted itself as a figure of Being, it has received the authority to do so only by

31 Matthew Holt, 'Catastrophe, autonomy and the future of modernism: Trying to understand Adorno's reading of *Endgame*', *Samuel Beckett Today – Aujourd'hui*, vol. 14 (2004), 260–75 (p. 264).

32 Holt, 'Catastrophe, autonomy and the future of modernism', p. 264.

33 Adorno, *Noten zur Literatur*, pp. 290–1; *Notes on Literature*, p. 244.

> asserting itself to be the fundamental characteristic of Being. But that is precisely what is false. The immediacy of individuation was deceptive; the carrier of individual experience is mediated, conditioned. *Endgame* assumes that the individual's claim to autonomy and being has lost its credibility. But although the prison of individuation is seen to be both prison and illusion – the stage set is the imago of this kind of insight – art cannot break the spell of a detached subjectivity; it can only give concrete form to solipsism. Here Beckett runs up against the antinomy of contemporary art.[34]

Adorno seems to see *Endgame* as the description of the collapse of bourgeois humanism. He does not use the word 'humanism' but in certain passages of this text he uses terms belonging to the same semantic sphere of 'humanism'. For example, when Adorno discusses Lukács' critique addressed to Beckett – i.e. Beckett ends up reducing humans to their animality, their 'animal qualities' [*ihre Tierheit*]'[35] –, he also notes: 'Differentiatedness [*Differenziertheit*], once the precondition of humanness [*Humanität*], is gradually becoming ideology'.[36] Here, against the backdrop of a catastrophe (*das Unheil*) whose name is uttered in silence, Adorno critically alludes to the traditional modern argument discourse that stresses the peculiarity of human beings, aiming to identify the specific characteristic that decisively marks human beings and distinguishes them from animals. But for Beckett, the human being [*Der Mensch*] is nothing more than what it has ultimately become.[37] To this Adorno adds:

> As in utopia, it is its last day that decides on the species. But mourning over this must reflect – in the spirit – the fact that mourning itself is no longer possible. No weeping melts the armor; the only face left is the one whose tears have dried up. This lies at the basis of an artistic method that is denounced as inhuman [*als inhuman*] by those whose humanness [*Menschlichkeit*] has already become an advertisement for

34 Adorno, *Noten zur Literatur*, pp. 299–300; *Notes on Literature*, p. 244.

35 Adorno, *Noten zur Literatur*, p. 289; *Notes on Literature*, p. 243.

36 'Was die Bedingung von Humanität war, Differenziertheit, gleitet in die Ideologie'. Adorno, *Noten zur Literatur*, p. 289; *Notes on Literature*, p. 243.

37 Adorno, *Noten zur Literatur*, p. 290; *Notes on literature*, p. 244.

> the inhuman [*fürs Unmenschliche*], even if they are not aware of it. Of the motives for Beckett's reductions of his characters to bestialized human beings [*auf den vertierten Menschen*], that is probably the most essential. Part of what is absurd in his writing is that it hides its face.[38]

Here, Adorno again alludes to Lukács' criticism of Beckett. Indeed, in 'The Ideology of Modernism' (German edition, 1957), Lukács considers Beckett as a pivotal figure expressing modernism, that modernism which is, in his opinion, politically debilitated, in particular insofar as it transforms mental illness from a deviant pathological state and a disagreement with prevailing ideas, norms, or authority, enabling challenges to the status quo, into a universal norm of human condition. Indeed, Lukács states: 'What served, with Musil, as the ideological basis of a new typology – escape into neurosis as a protest against the evils of society – becomes with other modernist writers an immutable *condition humaine*';[39] 'Beckett's Molloy is perhaps the *ne plus ultra* of this development [...]. He presents us with an image of the utmost human degradation – an idiot's vegetative existence'.[40] Nevertheless, according to Adorno, such a critique of Beckett that emphasises the non-human nature of his characters – their reduction to a vegetative or animal state – loses sight of Beckett's intentional denunciation of what human beings have become as a result of modern catastrophe, despite all the attempts by existentialism to save the individual, their freedom and their autonomy.

Indeed, *Endgame* reveals the individual 'to be a historical category, both the outcome of the capitalist process of alienation and a defiant protest against it, something transient himself. [...] The immediacy of individuation was deceptive; the carrier of individual experience is mediated, conditioned. *Endgame* assumes that the individual's claim to autonomy and being has lost its credibility'.[41] Overall,

38 Adorno, *Noten zur Literatur*, p. 290; *Notes on literature*, p. 244.

39 Georg Lukács, 'The Ideology of Modernism', in Lukács, *The Meaning of Contemporary Realism*, ed. and trans. by John Mander and Necke Mander (London: Merlin, 1963), pp. 17–46 (p. 31).

40 Lukács, 'The Ideology of Modernism', p. 31.

41 Adorno, *Noten zur Literatur*, pp. 290–1; *Notes on literature*, p. 244.

Adorno reads Beckett's text as a diagnosis of the collapse of modern (Cartesian) subjectivity: '*Endgame* is the epilogue [*Nachspiel*] to subjectivity, and the play loses the hero along with subjectivity. The only aspect of freedom still known to it is the powerless and pitiful reflex action of trivial decisions'.[42] In 'Commitment' he again stresses this aspect of Beckett' s theatre:

> Philosophical apologists may find it convenient to view Beckett's oeuvre as an anthropological sketch, but in fact it deals with an extremely concrete historical state of affairs: the dismantling of the subject. Beckett's *ecce homo* is what has become of human beings. They look mutely out from his sentences as though with eyes whose tears have dried up. The spell they cast and under which they stand is broken by being reflected in them. The minimal promise of happiness which they contain, which refuses to be traded for any consolation, was to be had only at the price of a thoroughgoing articulation, to the point of worldlessness.[43]

Emphasising that the subject's freedom consists of trivial decisions – as well as his rhetorical abilities in repetitions of banal or impotent questions and observations – *Endgame* expresses the end of bourgeois humanism, the quintessence of which is the celebration of freedom in the *Oratio* by Pico della Mirandola or, with other emphasis, in *Existentialism Is a Humanism* (1946) by Sartre. As Adorno writes in his notes on *Endgame*, '[t]he sovereign *ego cogitans* is transformed by the *dubitatio* into its opposite. And that is what it always was. For in order to retain itself as absolutely certain it had to turn itself into ever less'.[44] Moreover, in his notes and marginalia in Adorno's copy of the German translation of *The Unnamable*, Adorno detects in this novel 'the fate of the Cartesian

42 Adorno, *Noten zur Literatur*, p. 303; *Notes on literature*, p. 253.

43 Adorno, *Noten zur Literatur*, 'Engagement' pp. 409–30 (p. 425); *Notes on literature*, 'Commitment', pp. 348–63 (p. 360).

44 Theodor W. Adorno, 'Dossier: Adorno's Notes on Beckett', trans. by Dirk Van Hulle and Shane Weller, *Journal of Beckett Studies*, 19.2 (2010), 157–78 (p. 173).

subject as a movement towards "absolute alienation"'.[45] In Beckett's works, the power of reason and the modern subject is reversed into weakness and vulnerability:[46] it is limited to reflecting on its own failure and inaction. In *Malone Dies*, for example, at the beginning of the novel, the protagonist says: 'I will not weigh upon the balance any more, one way or another. I shall be neutral and inert. No difficulty there'.[47] He leads this remnant of existence in a state of doubt and loss of object references: 'I fear I must have fallen asleep again. In vain I grope, I cannot find my exercise-book. But I still have the pencil in my hand. I shall have to wait for day to break. God knows what I am going to do till then'.[48]

The discussion now returns to Adorno's reading of *Endgame*. He recalls the famous Jewish story, with metaphysical significance, told by Nagg, Hamm's father. After telling it, Nagg laughs but is silenced by Hamm.[49] And Adorno comments:

> The embarrassment that comes over us when someone laughs about his own words becomes existential; life is still a quintessence only as the quintessence of everything one has to be ashamed of. Subjectivity dismays us as domination in a situation where one person whistles and the other comes running. But what shame protests against has its social value: in the moments when the bourgeois act like true bourgeois, they sully the notion of humanity [*den Begriff der Humanität*] that is the basis for their own pretensions. Beckett's prototypes are also historical in that they hold up as typical of human beings only the deformations inflicted upon them by the form of their society. There is no room left for others. The bad habits and ticks of the normal personality, which

45 Shane Weller, 'Adorno's Notes on *The Unnamable*', *Journal of Beckett Studies*, 19.2, *Beckett and Germany* (2010), 179–195 (p. 183). See Theodor W. Adorno, *Frankfurter Adorno Blätter III*, ed. by Theodor W. Adorno Archiv (Munich: editiontext+kritik, 1994), p. 43.

46 See, for example, in Beckett's play, Hamm's parents, perceived as physically fragile, mortal, in a state of disintegration, and his repulsion towards medicine bottles. Adorno, *Noten zur Literatur*, p. 298; *Notes on literature*, p. 250.

47 Samuel Beckett, *Molloy, Malone Dies, The Unnamable* (London: Calder, 1994), p. 179.

48 Beckett, *Molloy, Malone Dies, The Unnamable*, p. 209.

49 Adorno, *Noten zur Literatur*, p. 299; *Notes on literature*, p. 251.

> *Endgame* intensifies unimaginably, are the universal form – which has long since put its stamp on all classes and individuals – of a totality that reproduces itself only in and through particularity in the bad sense, the antagonistic interests of individuals. But because there has been no life other than the false life, the catalog of its defects becomes the counterpart of ontology.[50]

It should be noted that in these pages Adorno repeatedly uses the words *Humanität* and *Menschlichkeit*. In the English version, both are translated by Shierry Weber Nicholson with 'Humaness'. In the French translation both are translated by Sibylle Müller with 'Humanisme'.[51]

As is well known, in German, 'Humanität', in accordance with its origin, covers three meanings: it can refer to *(a)* the human race (which is supposed to or likely to progress and evolve over the centuries); *(b)* the humanity of human beings (that which characterises them and distinguishes them from animals); and *(c)* the virtue of humanity (which moral subjects must demonstrate by showing others the respect they are due as human beings). *Menschlichkeit* can refer to the latter two meanings (that which characterises men and distinguishes them from animals, and virtue). The three concepts covered by *Humanität* are undoubtedly distinct, but they are also inseparable. All three meanings are constitutive of what Modernity has become: modern times affirm the existence of a human race that progresses by distinguishing itself ever more clearly from animals and that distinguishes itself by cultivating the virtue of humanity. The concept of *Humanität* would be thematized by Adorno in another essay of *Notes on literature*: 'On the Classicism of Goethe's *Iphigenie*'. Here he often uses *Humanität* to designate humanity in the sense of an achieved quality of humanness, in accordance with the Enlightenment ideal, as distinguished from the more generic *Menschheit* or *Menschlichkeit*. Specifically, here 'humanity' refers to the capacity for self-determination and emancipation of the

---

50 Adorno, *Noten zur Literatur*, pp. 299–300; *Notes on literature*, p. 251.

51 Theodor W. Adorno, *Notes sur la literature*, trans. by Sibylle Müller (Paris: Flammarion, 2009), pp. 209 and 210.

human individual.[52] But, Adorno states, in Iphigenie, *Humanität* arises from the experience of its antinomy: 'Once emancipated, the subject, which did not emancipate itself in the civilizing process so much as emerge from it, comes into conflict with civilization and its rules'. The emancipated subject – as Goethe shows in the tragedy of Torquato Tasso – 'cannot live freely in the bourgeois society that dangles freedom before it'.[53] Conversely, historically, *Humanität* emerges through a dialectic process, in which civilisation coexists with injustice.[54]

As already remarked, by interpreting Beckett, Adorno affirms that 'differentiatedness', the specific trait distinguishing humans from other beings, which was the precondition of humanness (*Humanität*), is gradually becoming ideology, i.e., it reveals itself to be untrue, a false construct that is passed off as solid, well-founded, eternal and universal, but rather contains a series of cracks that become increasingly evident over time. Conversely, appeals to *Menschlichkeit* have merely become an advertisement for the inhuman (*fürs Unmenschliche*). *Menschlichkeit* does not designate positive distinctive characteristics of man, but it is a clue to recognising the inhuman. Here, we find Adorno's conviction previously highlighted: the human cannot be defined; the inhuman, however can be identified. Consider his 1963 lecture *Problems of moral philosophy*:

52 Adorno, *Noten zur Literatur*, 'Zum Klassizismus von Goethes Iphigenie', pp. 495–514 (p. 499); *Notes on literature*, 'On the Classicism of Goethe's *Iphigenie*', pp. 415–29 (pp. 418–19).

53 Adorno, *Noten zur Literatur*, p. 500; *Notes on literature*, 'On the Classicism of Goethe's *Iphigenie*', p. 419.

54 'As long as this dialectical relationship could not yet be understood, the spokesperson for Humanität was forced to temper its civilizing moment with injustice. The latter, the residue of barbarism in the resistance to barbarism, is the surrogate for the reconciliation with nature that sheer opposition to myth did not succeed in bringing about. In Iphigenie injustice is done to those who are literally, in the Greek use of the term, barbarians [βάρβαροι or non-Greeks]. The barbarian nature of the non-Greeks is made crassly concrete in the custom, which Iphigenie suspends but does not abolish, of sacrificing a foreigner to the goddess'. Adorno, *Noten zur Literatur*, p. 507; *Notes on literature*, 'On the Classicism of Goethe's *Iphigenie*', p. 424.

> We may not know what absolute good is or absolute norm, we may not even know what man is or what the human or humanity [*der Mensch or das Menschliche und die Humanität ist*] – but what the inhuman is we know very well indeed.[55]

The passages examined reveal that, in Adorno's interpretation of *Endgame*, bourgeois society dishonours and contradicts the values of its own presumed humanism because individuals suffer from cognitive distortions – which manifest themselves in their tics and idiosyncrasies – produced by the form of their society. Brutality dominates human actions (Hamm does nothing to satisfy the hunger of his parents, reduced to a childlike state in the rubbish bins) and what poor humanity remains (the parents sharing the last biscuit) appears repugnant. What remains of love – since the two cannot even embrace each other – is reduced to the pure act of swallowing and its noises.[56]

In such a situation, the possibility of a positive and authentic relationship between human beings and nature is precluded. The situation in the play is none other than that in which 'there's no more nature [*keine Natur mehr gibt*]'. 'The phase of complete reification of the world [*der vollendeten Verdinglichung*]' – Adorno writes –, where there is nothing left that has not been made by human beings, is indistinguishable from an additional catastrophic event caused by human beings, in which nature has been wiped out and after which nothing grows any more'.[57] In Beckett's one-act play, Hamm evokes nature explaining that it has abandoned them, while Clov replies that there is no more nature:

> HAMM: Nature has forgotten us.
> CLOV: There's no more nature.
> HAMM: No more nature! You exaggerate.
> CLOV: In the vicinity.
> HAMM: But we breathe, we change! We lose our hair, our teeth! Our bloom! Our ideals!

---

55 Adorno, *Probleme der Moralphilosophie*, p. 261; *Problems of moral philosophy*, p. 175.

56 Adorno, *Noten zur Literatur*, pp. 301–2; *Notes on literature*, pp. 252–3.

57 Adorno, *Noten zur Literatur*, pp. 285–6; *Notes on literature*, pp. 240–1.

CLOV: Then she hasn't forgotten us.
HAMM: But you say there is none.
CLOV (*sadly*): No one that ever lived ever thought so crooked as we.
HAMM: We do what we can.
CLOV: We shouldn't. (*Pause.*)[58]

While this exchange and the following pages evoke the spectre of a nature that, in the age of catastrophe, no longer produces what humans need, later Beckett's text highlights the profound indifference of the subject towards the universe, the entire reality outside him.[59] In any case, it is interesting to note that Adorno considers the possibility that, driven by their boundless instinct for survival, human beings will destroy the natural world along with themselves. Whether everyone's death is the result of a nuclear war waged in self-defence, or the pillage of the Earth in the name of profit, is not really decisive. Both scenarios are equally possible; indeed, the second prospect is becoming progressively more plausible. The refusal of the owners of the means of production to consider these possibilities, as well as the general and voluntary subjugation to capital, make the end of life on this planet an all-too-plausible scenario.[60]

The final balance is a sort of unveiling and dialectic of reason that reveals the lack of meaning both of the subject and nature: reason itself terminates in nonsense. In one of the last pages of

58 'HAMM. – La nature nous a oublié.
CLOV. – Il n'y a plus de nature.
HAMM. – Plus de nature! Tu vas fort.
CLOV. – Dans les environs.
HAMM. – Mais nous respirons, nous changeons! Nous perdons nos cheveux, nos dents! Notre fraîcheur! Nos ideaux!
CLOV. – Alors elle ne nous a pas oublié.
HAMM. – Mais tu me dis qu'il n'y en a plus.
CLOV (*tristement*). – Personne au monde n'a jamais pensé aussi tordu que nous.
HAMM. – On fati ce qu'on peut.
CLOV. – On a tort.
Un temps'. Beckett, *Fin de partie*, pp. 23–4.

59 'HAMM (*angrily*): To hell with the universe [HAMM (*avec colère*). Je m'en fous de l'universe'. Beckett, *Fin de partie*, p. 63.

60 On this point, see Cook, Adorno on Nature, pp. 102–3.

Adorno's essay, as noted above, he states: 'No statement sounds more absurd than this most rational of statements, which reduces "everything" to an "only", the mirage of a world that can be dominated anthropocentrically [*dem Trugbild der anthropozentrisch beherrschbaren Welt*]'.[61] A few lines later, Adorno affirms:

> Ratio, which has become completely instrumental [*vollends instrumentell geworden*], devoid of self-reflection and reflection on what it has disqualified, must inquire after the meaning that it itself has expunged. But in the state that makes this question necessary there is no answer left but the void that the question, as pure form, already is. The historical inevitability of this absurdity makes it seem ontological: that is the delusoriness of history itself. Beckett's drama demolishes it. The immanent contradiction of the absurd, the nonsense in which reason terminates [*der Ursinn, in dem Vernunft terminiert*], opens up the emphatic possibility of something true that cannot even be conceived of anymore. It undermines the absolute claim of the status quo, that which simply is the way it is. Negative ontology is the negation of ontology: it was history alone that produced what the mythical power of the timeless and eternal has appropriated.[62]

In this passage, Adorno refers to reason that is no longer capable of reflecting on itself, a theme that, as argued earlier, was already addressed in *Dialectic of Enlightenment*.[63] This purely instrumental reason is also devoid of reflection on what it has disqualified, namely nature and non-human living beings. Beckett's theatre reveals that purely instrumental reason leads to absurdity, to nonsense, but it is precisely nonsense – as the opposite of purely instrumental reason – that can give us a glimpse of the possibility of a truth different from that which is considered to be true and reliable in the sphere of domination. From this perspective, negative ontology as the negation of ontology heralds the approach of negative dialectics, which theoretically highlights the limits of the subject and the role of the object that transcends the concept. It concerns both nature and human beings. In the case of nature, this

61 Adorno, *Noten zur Literatur*, p. 318–19; *Notes on literature*, p. 265.

62 Adorno, *Noten zur Literatur*, p. 319; *Notes on literature*, p. 265.

63 Horkheimer and Adorno, *Dialektik der Aufklärung*, pp. 18–19 and 30; *Dialectic of Enlightenment*, p. xvi and p. 4.

approach allows us to recognise the preponderance of the object over the concept, and thus to avoid the violence of the concept against the object and, therefore, on nature. From this perspective, nature – more precisely the natural and biological framework – is always socio-historically shaped; but as it is the material basis of human life and not merely an image or model, material nature is irreducible to any single social configuration.[64] In *Negative Dialectics*, Adorno stresses the 'preponderance' of the object over the mind due to the fact that matter (*Stoff*, *Materie*) prevails over spirit. Human beings can grasp this preponderance subjectively by reflecting on their experience.[65] Nevertheless, Adorno also complains that, when a thing becomes an object of cognition, 'its physical side is spiritualized [*vergeistigt*] from the outset by translation into epistemology'.[66] Adorno calls into question and rejects such a spiritualization, in order to do justice to things by unveiling those aspects of them that are not identical with – and thus irreducible to – concepts. These 'nonidentical moments show up as matter, or as inseparably fused with material things'.[67] Specifically, Adorno advocates a 'disenchantment of concept'. Indeed, as he states in the 'Introduction' of *Negative Dialectics*, 'all concepts, even the philosophical ones, refer to nonconceptualities [*Nichtbegriffliches*], because concepts on their part are moments of the reality that requires their formation, primarily for the control of nature'.[68] While, on the one hand, concepts refer to non-conceptual, material particulars, on the other hand, they are shaped in historically situated and conditioned encounters with them. Concepts are 'entwined with a nonconceptual whole': its characteristic is 'to refer to nonconceptualities';[69] in other words, what survives in concept by dint of its meaning is its 'nonconceptual conveyance or transmission (*Vermitteltsein*) under specific historical conditions'.[70]

---

64 See Nelson, 'Revisiting the Dialectic of Environment', p. 124.
65 Adorno, *Negative Dialektik*, p. 186; *Negative Dialectics*, p. 185.
66 Adorno, *Negative Dialektik*, p. 193; *Negative Dialectics*, p. 192.
67 Adorno, *Negative Dialektik*, p. 193; *Negative Dialectics*, p. 192. On this point, see Cook, *Adorno on Nature*, p. 10.
68 Adorno, *Negative Dialektik*, p. 23; *Negative Dialectics*, p. 11.
69 Adorno, *Negative Dialektik*, p. 24; *Negative Dialectics*, p. 12.
70 Cook, *Adorno on Nature*, pp. 10–11.

It seems reasonable to conclude this paragraph by returning to the concept of *Humanität* and referring to its use in some of Horkheimer's post-war writings. The word *Humanität* is the title of an aphorism – written between 1956 and 1958 – in Horkheimer's work *Notizen*. Here he writes:

> Our practical philosophy is humanity [*die Humanität*]. That people [*Das Menschen*] should not suffer any hardship [*kein Elend leiden*], or rather, indeed that creature may unfold, is the purpose of general action [*ja daß Kreatur sich entfalten dürfe, ist der Zweck das allgemeinen Handelns*]. [...] Admittedly, this sounds banal [*kitsching*], but we know no other criteria of equal rank.[71]

Once again, as he did in *Dämmerung*, Horkheimer affirms that *die Humanität* consists in fighting against any suffering. Once again, he suggests that suffering unites and equates human beings and animals. Nevertheless, in this note he considers the idea of a practical humanism which promotes the blooming and development of each creature, almost evoking the idea of a mutual care in nature[72] which should replace the folly and cruelty of man towards animals, mentioned once again in *Notizen*, particularly in the aphorism 'Mensch und Rhinozeros'.[73] This 'humanism', this practical philosophy under the banner of humanity, seems to echo

71 Horkheimer, *Notizen 1949-1969*, pp. 259–60; *Decline: Notes 1950-1969*, p. 152 (translation amended).

72 It is an idea that Horkheimer was able to find in Montaigne. 'Even if all this were lacking, still there is a certain respect, and a general duty of humanity, that attaches us not only to animals, who have life and feeling, but even to trees and plants. We owe justice to men, and mercy and kindness to other creatures that may be capable of receiving it. There is some relationship between them and us, and some mutual obligation'. *The Complete Essays of Michel de Montaigne*, trans. by Donald M. Frame (Stanford: Stanford University Press, 1958), II:11, 'Of cruelty', p. 318.

73 'Humans are the only species that keeps members of other species captive, tormenting them in various ways, for the sole purpose of feeling superior'. Horkheimer, *Notizen 1949-1969*, p. 266. In this aphorism, Horkheimer evokes an eighteenth-century painting by Longhi depicting a captive rhinoceros, admired by the elegant public of a European city. It is interesting to note that the figure of the rhinoceros often recurs in Adorno's

the negative humanism to which Horkheimer refers in *Späne*, in the 1952 aphorism 'Negativer Humanismus'.[74] However, the objectives, principles and content of this humanism that can be adequately described (essentially also in a negative way) appear 'kitsch', even though they are indispensable. This indicates, however, not only that traditional forms of humanism are now obsolete and outdated (and even 'disreputable and meaningless', as some might say[75]); the connotation of 'kitsch' would also evoke the reduction of humanism to slogans, to pure rhetoric, which serves as a theoretical smokescreen for inhumane social practices.[76]

Nevertheless, compared to Adorno, Horkheimer – at the end of the 1950s – seems to believe that it is possible to go beyond a 'negative dialectic'. Of course, he too believed that it was not possible to determine positively what it is to be human, partly because he always rejected any discourse on essences that transcend concrete historical epochs, but he thought it possible to circumscribe the characteristics of an active humanism. This humanism, by recording suffering and inhumanity in reality, can lay the foundations for a society that combats the exploitation of human beings, as well as the exploitation of nature, and allows every living being to develop without being exposed to unnecessary suffering. Indeed, in a note dating from 1957–58, Horkheimer states: '[Marx] […] imagined that peace among the classes would also be peace among men and with nature. […] Marx would probably really have believed that a liberated mankind [*die freie Menschheit*] would send rockets to the moon – because it was curious or pour passer le temps. But it turns out that rockets are part of the realm of necessity. What remains to

texts, for example in *Ästhetische Theorie*, p. 171–2; *Aesthetic Theory*, p. 112. See § 3 of this chapter.

74 See chapter II, § 2 of our book.

75 Horkheimer, *Dämmerung*, 'Entehrte Begriffe' p. 315; *Dawn: Notes 1926-1931*, 'Debased Concepts', p. 18.

76 See Martin Desrosiers, 'Un humanisme négatif. Horkheimer, Adorno et la critique du "kitsch humaniste"', in *Les Normes et le possible. Héritage et perspectives de l'École de Francfort*, ed. by Pierre-François Noppen, Gérard Raulet and Iain Macdonald (Paris: Éditions de la Maison des sciences de l'homme, 2012), pp. 285–307 (p. 286).

freedom is solidarity with life, the struggle for what is right, not only in society but in nature at large'.[77]

3. *Humanism in Adorno: 'Kultur und Culture' and 'Über Technik und Humanismus'*

In the preface of the 1969 edition of *Dialectic of Enlightenment*, Horkheimer and Adorno write:

> In a period of political division into immense blocs driven by an objective tendency to collide, horror has been prolonged. The conflicts in the third world and the renewed growth of totalitarianism are not mere historical interludes any more than, according to the *Dialectic*, fascism was at that time. Critical thought, which does not call a halt before progress itself, requires us to take up the cause of the remnants of freedom, of tendencies toward real humanity [*für Tendenzen zur realen Humanität*], even though they seem powerless in face of the great historical trend.[78]

Here Horkheimer and Adorno still evoke the idea of true humanity and the possibility of identifying trends towards overcoming the inhuman condition of present-day societies.

It is now necessary to examine a lecture given by Adorno in 1957, which may perhaps help us shed light on the passage from the 1969 Preface to *Dialectic of Enlightenment*. It is a lecture given at the Münchner Amerika-Haus on 7 June 1957, entitled: 'Sind amerikanische und deutsche Kultur vergleichbar?' Its transcript was published under the title 'Kultur und Culture'. The following passage from this text merits emphasis:

> We do not actually become human beings [*eigentlich Menschen*] by fulfilling ourselves as distinct individuals [*als je einzelnen*], but by stepping out of our shells, by entering into relations with other

77 Horkheimer, *Notizen 1949-1969*, 'Die drei Fehler von Marx' pp. 269–70; *Decline: Notes 1950-1969*, 'The Three Mistakes of Marx', pp. 156–7.

78 Horkheimer and Adorno, *Dialektik der Aufklärung*, p. 13; *Dialectic of Enlightenment*, p. xi.

> people [*in diesem Aus-uns-Herausghen zu andern Menschen*], and by – in a certain sense – abandoning ourselves to them. Only by going through such externalization [*Veräußerlichung oder Entäußerung*] do we determine ourselves as individuals, and not by – as Wilhelm von Humboldt's concept of *Bildung*, for instance, expected of us – pouring water over ourselves as if we were plants in order to become comprehensively educated personalities.[79]

Commenting this passage, Kozlarek writes: 'Humanism, then, is neither a philosophical theory, nor a clearly defined educational program, but a kind of social and cultural practice'.[80] According to Kozlarek, through the concept of 'real humanity' Adorno would contrast 'an overly spiritualized humanism' which is based on a limited and incomplete view of human beings: 'the individual enclosed within himself'.[81]

Nevertheless, Adorno's lecture also addresses – decisively – the issue that concerns us: the connection between humanism and nature: how nature, non-human living beings and environmental resources can be conceived through our conceptual tools, in order to rethink and redefine the relationship between human communities and the natural environment – differently from the domination and indiscriminate exploitation of natural resources.

Here Adorno identifies two typical meanings of the word 'culture' that are linked to the opposition between American culture and German culture, which is the subject of the lecture.

> On the one hand, culture can, in fact, refer to the human being's coping with [*die Bewältigung*] nature in the sense of its mastery [*ihrer Beherrschung*]; that is, domination [*der Herrschaft*] of both the external nature that opposes us as well as domination of the natural forces in the

79 Theodor W. Adorno, 'Kultur und Culture' [7.6.1957], *Vorträge 1949-1968*, ed. by Michael Schwarz, *Nachgelassene Schriften*, Abteilung V: *Vorträge und Gespräche*, Vol. 1 (Berlin: Suhrkamp Verlag, 2019), pp. 156–76 (p. 165); 'Kultur and Culture', *Social Text*, 27.2/99 (2009), 145–158 (p. 152).

80 Oliver Kozlarek, *The Critical Humanism of the Frankfurt School as Social Critique* (Lanham, Maryland: Lexington Books, 2024), p. 125.

81 Kozlarek, *The Critical Humanism of the Frankfurt School as Social Critique*, p. 125.

> human being itself – in a word: the control of civilization over human urges and the unconscious. One could characterize such a notion of culture as one whose substance is essentially the molding of reality. I thereby do not want to say that in America reality itself has in a radical sense been shaped, that it has in a positive sense been stripped of its organic nature.[82]

Bewältigung (*/Überwältigung*) contains the idea of management, control, but literally it would mean 'to get the better of'. *Beherrschung* and *Herrschaft* designate respectively the process and its result; they can be translated by 'subjugation' and 'domination'. Moreover, in the writings of Horkheimer and Adorno, *Herrschaft* refers to the concrete and massive power of human beings over nature – almost a sovereignty of the human being over the natural world as the result of the process of subjugation.

Consider now what Adorno writes in the second part of this passage ('On the one hand, "culture" can mean …'). Here, he specifies that the heart of the contrast between American and German culture rests on two different approaches towards the word *culture*: *(1)* gaining mastery over one's natural surroundings and human nature; and *(2)* caring for and preserving nature that the human power simultaneously destroys.

> This moment, however, is now confronted by another aspect of the concept of culture. To care for something [*Pflegen*] is not simply the same thing as to dominate it [*wie herrschen*]. Tending nature is not simply [*Die Natur pflegen heißt nicht einfach*] the same thing as suppressing or exploiting it [*die Natur unterdrücken und die Natur ausbeuten*]; rather, the notion of care [*Pflege*] also contains the moment of conserving [*das Moment des Bewahrens*] something for its own sake, the idea that what human beings acquire [*was von den Menschen angeeignet*] and subject to their rule [*was unter ihre Herrschaft gebracht wird*] should not be radically broken or eradicated but rather simultaneously preserved in its own being. In order to illustrate this idea of culture, I would like to ask you to, for instance, think of a very good wine, an artificial product, which has been distilled from grapes in a complicated fermentation process. Artificial as it may be, one can nevertheless still taste the earth in which it grew or the grapes from which it was

82 Adorno, 'Kultur und Culture', p. 156; 'Kultur and Culture', p. 146.

> pressed. Think of certain fine *vins du pays*, especially French ones. Such wines are specific products of culture; and this notion of culture, that is, the idea of preserving that which in nature is simultaneously destroyed [*aufgehoben ist*] by human power [*menschlicher Kraft*], is what the European conception of culture is to a large extent based on. This, however, now also entails a certain amount of mindfulness [*des Eingedenkens*] and self-reflection, a moment of internalization if you will, which breaks the immediate power over nature. In Germany in particular, and perhaps more so in Germany than in any other country in the world, this notion of culture results from what in a concise sense might be called *Geisteskultur*. That is, a culture that – to illustrate this conception explicitly – essentially manifests itself in what are called great works – be they of art, philosophy, or scholarship.[83]

This passage is, from this perspective, particularly eloquent: it shows, once again, the intertwining of human activities (human power) and natural resources and elements, that is, the difficulty of separating them in a clear and definitive way, so as to be able to identify and designate a clear essence both of human being and nature in its purity, separate from human activity; in other words, the illusion of capturing nature in its untouched state. Not surprisingly, the term 'culture', in its original classical meaning, refers to the relationship that human beings establish with the environment and natural resources thanks to their intelligence and technical skills.

Conversely, at the present time the notion that nature and culture cannot be conceived in isolation from one another is shared by various authors who have radically rethought or are thoroughly rethinking the connections between humans, non-human animals and the environment. These authors consider the notion that humans exist to be almost self-evident because of their embeddedness within networks of associations between humans and nonhumans: it is ceaselessly disclosed by their experience of the world. Despite this, these authors believe that there has been a failure to develop a dynamic notion of 'a common world' capable of guiding human modes of thinking and shaping collective activities. As Tomas Weber states, humans have not internalised the notion of a common world to such an extent that they are able 'to actively pursue, foster, and care

83 Adorno, 'Kultur und Culture', p. 157; 'Kultur and Culture', pp. 146–7.

for it'.[84] In *Facing Gaia*, for example, Latour affirms that 'nature' and 'culture' should be considered not in terms of opposition, but in terms of relationship within a broader framework in which nature and culture constitute a special case.[85] However, Latour tends to suppress the concept of nature, or rather to discard '*the ideology of "nature"*',[86] while Adorno does not question the concept of nature itself, but emphasises that we do not have access to nature as such: it is always mediated by society. From this perspective, Adorno's viewpoint can also open up a dialogue – perhaps even more so – with Latour's critics, who continue to use the notion of nature.[87] Specifically, by preserving the concept of nature, a distinction between what depends on human responsibility and what proceeds from the rest of nature can be drawn; moreover, it is possible to acknowledge – as emerges in the last passage quoted from Adorno – the possibility of a relationship between human beings and nature based on 'care' and 'caring' rather than domination and exploitation.

The analysis now moves to another text by Adorno from the 1950s, which is believed to be the only one of Adorno's corpus in which the word 'humanism' appears in the title. In this text, entitled 'On Technology and Humanism' ('*Über Technik und Humanismus*'), and which was preceded by a lecture on 10 November 1953, given at the Technical University of Karlsruhe, Adorno outlines his conception

---

84 Tomas Weber, 'Metaphysics of the Common World: Whitehead, Latour, and the Modes of Existence', *The Journal of Speculative Philosophy*, 30.4 (2016), 515–33 (p. 515).

85 'We would have to be able to introduce an opposition, not between nature and culture this time (since the incessant vibrations between the two are what drives us crazy), but between Nature/Culture on one side and, on the other, a term that would include each one of them as a particular case. I propose simply to use the term *world*, or "worlding," for this more open concept, defining it, in an obviously very speculative fashion, as that which opens to the multiplicity of *existents*, on the one hand, and to the *multiplicity* of ways they have of existing, on the other'. Latour, *Face à Gaïa*, p. 49; *Facing Gaia*, p. 35.

86 Latour, *Down to Earth*, p. 65.

87 For these discussions, see Fabrice Flipo, *Les écologistes sans la nature. Le faux ami Bruno Latour* (Paris: Textuel, 2025), pp. 94–8.

of humanism by distinguishing it from technology.[88] In this short text he affirms that the antithesis of humanism and technology is artificially produced in a society fragmented in distinct domains following distinct rationalities.[89] Conversely, he states that learning is not adaptation, even though it is inconceivable without a moment of adaptation, just as it is inconceivable without a substance that transforms and persists. Adorno points out that culture is not something more elevated and refined than technology.[90] However, he clearly presents technology, not as a great entity endowed with autonomous life, but as a product historically derived from human activity, and therefore attributes responsibility for it to the human being.

> But technology [*die Technik*] is not the primary social being, not the thing itself, not humanity [*die Menschheit*], but only something derived from it, the organisational form of human labour. I consider the conceptualisation of a new educational ideal, such as the synthesis of humanistic and technical man, to be as unfruitful as all cultural syntheses. It is a commonplace, which only superstition in the omnipotence of measures compels us to repeat, that education cannot be decreed. It must emerge from objective historical conditions.[91]

In reality, in this text Adorno emphasises the importance of the critique of education, intertwined with a critical awareness of technology[92] as well as the understanding of the social contexts in

---

88 See Kozlarek, *The Critical Humanism of the Frankfurt School as Social Critique*, p. 136.

89 Theodor W. Adorno, 'Über Technik und Humanismus', in Adorno, *Vermischte Schriften I. Theorien und Theoretiker – Gesellschaft, Unterricht, Politik*, *GS* 20.1, 5th edition (Frankfurt am Main: Suhrkamp, 2023), pp. 310–17 (p. 314).

90 Adorno, 'Über Technik und Humanismus', p. 316.

91 Adorno, 'Über Technik und Humanismus', pp. 316–17.

92 Fifteen years later, in *Introduction to Sociology* (1968), the last lecture he ever gave, Adorno would emphasise the connection between technology, domination over nature and domination over human beings: 'the very concept of technology contains an ineradicable moment of dominance over nature, and as soon as this idea of technology as dominance over nature is applied directly and without reflection to human beings, the

which humans are entangled: the hope for a form of education that does not aim merely at intellectual self-development and the cult of personality, lies in this critical attitude.[93] Indeed, social criticism, opposing an idea of humanism as a domain detached from other social spheres, emphasises the need to rethink humanism based on the actual material conditions in which education and culture take shape. In other words, according to Adorno, reality, on which the development of the individual relies, is a historical social totality, and not an ahistorical realm of ideas and values or culture untouched by the material, economic, and political production and reproduction of society.[94]

---

notion of dominance over nature is transferred just as directly to human beings'. Theodor W. Adorno, *Introduction to Sociology*, ed. by Christoph Godde, trans. by Edmund Jephcott (Stanford, CA: Stanford University Press, 2000), p. 135.

93 'Heute gibt nur in der Kritik der Bildung und andererseits im kritischen Bewußtsein der Technik von sich selber und der Einsicht in die gesellschaftlichen Zusammenhänge, in die wir verstrickt sind, die Hoffnung auf eine Gestalt der Bildung sich kund, die sich nicht mehr die Humboldtische, längst schattenhafte Pflege der Persönlichkeit zur Aufgabe setzt'. Adorno, 'Über Technik und Humanismus', p. 317. Here Adorno evokes Humboldt's elitist notion of the educated personality. Undoubtedly, Adorno considers this excessively individualistic and subjectivist element to be a problematic aspect of Humboldt's pedagogy. Nevertheless, as Aurélia Peyrical points out, he finds in his works 'a strong emphasis upon the dialectic between the subject (the individual, *Einzelne*) and the object (the world, *Wirklichkeit*)', a dialectic embodied in the concept of *Entäusserung* (externalization). Aurélia Peyrical, 'T. W. Adorno, H. Becker, and the Challenges of Education in an "Administered World" (1955–1969): Unpublished Radio Conversations from the Theodor W. Adorno Archive', in *The Archives of Critical Theory*, ed. by Isabelle Aubert and Marcos Nobre (Cham, Switzerland: Springer, 2023), pp. 165–89 (p. 183). On the pedagogical dimensions of Adorno's work, see particularly Theodor W. Adorno, *Erziehung zur Mündigkeit: Vorträge und Gespräche mit Hellmut Becker 1959-1969*, ed. by Gerd Kadelbach (Frankfurt am Main: Suhrkamp, 1971); Theodor W. Adorno & Hellmut Becker, 'Education for maturity and responsibility', *History of the Human Sciences* 12.3 (1999), 21–34.

94 See Peyrical, 'T. W. Adorno, H. Becker, and the Challenges of Education in an "Administered World" (1955–1969)', p. 183; Kozlarek, *The Critical Humanism of the Frankfurt School as Social Critique*, p. 137.

In conclusion, in seeking to highlight the contents of humanism in an era dominated by technology, Adorno once again emphasises the need to take into account the inhumanity embedded in our societies:

> The form in which we can experience real humanism here and now, is the incorruptibility of thought [*die Unbestechlichkeit des Gedankens*] and fearlessness in the face of inhumanity [*die Unerschrockenheit im Angesicht der Unmenschlichkeit*], which does not come from technology, nor from individual people, but from the fatality [*von der Fatalität*] of what we are all involved in, every human being, all over the world.[95]

This fatality refers to the pervasiveness and omnipresence of the capitalist process of production as well as the mechanism of the administered world governed by instrumental reason, on which human lives now completely depend.[96] *Negative Dialectics* acts as a counterpoint to this passage and explores its contents in greater depth in the paragraph entitled 'Incapacitation of the subject'. Here Adorno observes the crisis of 'the anthropocentric sense of life [*das anthropozentrische Lebensgefühl*]', stemming from the foundation of idealism, from the very domination of nature, which has revealed the finitude of human beings in relation to their overly high expectations and their entanglement in an irrational socio-economic order.[97]

> There is a universal feeling, a universal fear, that our progress in controlling nature [*Naturbeherrschung*] may increasingly help to weave the very calamity [*an dem Unheil*] it is supposed to protect us from, that

95 'Die Form, in der wir heute und hier allenfalls den Humanismus real erfahren können, ist die Unbestechlichkeit des Gedankens und die Unerschrockenheit im Angesicht der Unmenschlichkeit, die nicht von der Technik herrührt und nicht von den einzelnen Menschen, sondern von der Fatalität dessen, worin wir alle, ein jeder Mensch, auf der ganzen Welt eingespannt sind'. Adorno, 'Über Technik und Humanismus', p. 317.

96 On the notion of 'administered world', see Theodor W. Adorno, Max Horkheimer, and Eugen Kogon, 'Die verwaltete Welt oder: Die Krisis des Individuums' [1950], in Horkheimer, *GS* 13: *Nachgelassene Schriften 1949–1972*, ed. by Gunzelin Schmid Noerr (Frankfurt am Main: Fischer, 1989), pp. 122–42.

97 Adorno, *Negative Dialektik*, p. 75; *Negative Dialectics*, p. 67.

> it may be weaving that second nature into which society has rankly grown. Ontology and the philosophy of Being are modes of reaction in which – along with other and cruder modes – consciousness hopes to escape from that entanglement. But they contain a fatal dialectics [*eine fatale Dialektik*]. The truth that expels man from the center of creation and reminds him of his impotence – this same truth will, as a subjective mode of conduct, confirm the sense of impotence, cause men to identify with it, and thus reinforce the spell of the second nature.[98]

The fatal dialectics in which human beings are involved arises from the crisis of anthropocentric humanism, caused in turn by the very claim to know nature in order to dominate it. The marginalisation of human beings on an ontological level ends up reinforcing their sense of powerlessness and increasingly crystallising them in the reified world of second nature. Adorno, therefore, regards doctrines that convey belief in being and that undermine the role of the subject, as harmful:

> Doctrines which heedlessly run off from the subject to the universe, along with the philosophy of Being, are more easily brought into accord with the world's hardened condition and with the chances of success in it than is the tiniest bit of self-reflection by a subject pondering upon itself and its real captivity.[99]

Here, Adorno openly argues with Heidegger. A little further on, he refers just as explicitly to Heidegger's critique of humanism, wondering whether the author of *Being and Time* does not intend to abolish the 'current talk of humanism [*das Gerede vom Humanismus*]', which is in itself despicable, with a view to destroying its substance.[100]

From the introduction to *Negative Dialectics* onwards, Adorno demonstrates the problematic nature of a hierarchical model of human-nature relations in which the human subject occupies a superior position with respect to natural entities or, at least, a central position in the cosmos. In this scheme, the relationship between the

98 Adorno, *Negative Dialektik*, pp. 75–6; *Negative Dialectics*, pp. 67–8.
99 Adorno, *Negative Dialektik*, p. 76; *Negative Dialectics*, p. 68.
100 Adorno, *Negative Dialektik*, p. 96; *Negative Dialectics*, p. 89.

individual and nature is characterised by separation. Human beings seek to separate themselves from nature, denying the cognitive roots in the animal sphere and the very primacy of the law of self-preservation, which contrast with the model of the *zoon politikon* that they have fabricated.[101] Living animals are characterised as evil and nature as a whole as inferior: this justifies every action of human beings aimed at dominating other living beings. Philosophical tradition has conceptually sanctioned this asymmetry and human prerogatives:

> The animal to be devoured must be evil. The sublimation of this anthropological schema extends all the way to epistemology. Idealism – most explicitly Fichte – gives unconscious sway to the ideology that the not-I, *l'autrui*, and finally all that reminds us of nature is inferior, so the unity of the self-preserving thought may devour it without misgivings. This justifies the principle of the thought as much as it increases the appetite. The system is the belly turned mind, and rage is the mark of each and every idealism. It disfigures even Kant's humanism [*Humanität*] and refutes the aura of higher and nobler things in which he knew how to garb it.[102]

Paradoxically, any attempt by human beings to separate themselves from and oppose nature ends in their own objectification. As the passage quoted above clearly illustrates,[103] the victims of anthropocentrism are both nature and humanity. Nevertheless, by declaring nature to be inferior to itself, the subject treats itself as a thing. Not by chance, *Dialectic of Enlightenment* aimed to demonstrate that the logic of self-preservation is the link between contempt for nature and the self-objectification of subjectivity.[104] What constitutes the centre of modern, enlightened anthropocentrism is not, in fact, the concrete human beings with their material dignity

101 Adorno, *Negative Dialektik*, p. 33; *Negative Dialectics*, p. 22.
102 Adorno, *Negative Dialektik*, pp. 33–4; *Negative Dialectics*, pp. 22–23.
103 See Adorno, *Negative Dialektik*, pp. 75–6; *Negative Dialectics*, pp. 67–8.
104 See Andrés Felipe Parra-Ayala, 'Materialismo ecológico como materialismo de la reproducción: un diálogo entre Theodor Adorno y Raquel Gutiérrez Aguilar', *Revista de Estudios Sociales*, 91 (2025), 23–38 (p. 27).

and needs, but the process of capital accumulation and valorisation. It might seem, then, that the use of the term anthropocentrism to refer to the modern, enlightened relationship with nature ends up obscuring the problem that what is at the centre of the human predatory relationship with nature is capital and not the human being. Moreover, according to Adorno, the consolidation of the capitalist relationship with nature goes hand in hand with the 'fallacy of constitutive subjectivity', that is the ideological self-image of the subject – of thought and self-consciousness – as the absolute sources of meaning and objectivity in reality: 'To use the strength of the subject to break through the fallacy of constitutive subjectivity – this is what the author felt to be his task ever since he came to trust his own mental impulses; now he did not wish to put it off any longer. Stringently to transcend the official separation of pure philosophy and the substantive or formally scientific realm was one of his determining motives'.[105] As Andrés Felipe Parra-Ayala has stated, '[a]lthough the anthropocentric conception of the world is sarcastically realised in the world of commodity fetishism – whose centre is not people, but always things – this sarcastic realisation is its necessary realisation. Anthropocentrism is the manifestation of the process of capital valorisation and accumulation in the image that human beings have of themselves and the surrounding world'.[106] Anthropocentrism is also evident in the way humans view and present animals, particularly those from exotic countries (which, incidentally, highlights the link between the exploitation of nature and colonialism). In *Aesthetic Theory*, for example, Adorno writes: 'That aspect of the Etruscan vases that most resembles speech depends most likely on their *Here I am* or *This is what I am*, a selfhood not first excised by identificatory thought from the interdependence of entities. Thus the rhinoceros, that mute animal, seems to say: "I am a rhinoceros"'.[107] 'I am a rhinoceros' indicates

105 Adorno, *Negative Dialektik*, p. 10; *Negative Dialectics*, p. xx.

106 Parra-Ayala, 'Materialismo ecológico como materialismo de la reproducción', p. 25 (translation by the author).

107 On this subject, see the collective volume *Adorno's Rhinoceros: Art, Nature, Critique*, edited by Antonia Hofstätter and Daniel Steuer (London: Bloomsbury, 2022). In particular, on the comparison between the

the possibility of a society in which the word 'animal' is no longer an insult nor immediately brings to mind an entity to be subjugated and exploited. As Camilla Flodin affirms, according to Adorno, 'the revaluation of nonhuman animals is a necessary condition for the possibility of the human animal to truly come into her own'.[108]

In the light of all these considerations, it can be said that 'real humanism' does not require the decentralisation of the subject, but rather its ability to reflect on itself and its condition in the administered world. Moreover, it relies on 'a universal feeling, a universal fear' about the large-scale consequences of domination over nature, which include the possibility of 'the annihilation of all life on the planet'.[109] From this perspective, even the natural beauty – apparently moving away from a view of nature as merely a collection of objects to be dominated and exploited – is involved in the reified and administered world. As Adorno writes in *Aesthetic Theory*,

> for in every particular aesthetic experience of nature the social whole is lodged. Society not only provides the schemata of perception but peremptorily determines what nature means through contrast and similarity. Experience of nature is coconstituted by the capacity of determinate negation. With the expansion of technique and, even more important, the total expansion of the exchange principle, natural beauty increasingly fulfils a contrasting function and is thus integrated into the reified world it opposes.[110]

Therefore, a thematization of what natural beauty has become in western society can fuel self-reflection in subjects who question their own condition in a world subject to the logic of planning, rationalisation and bureaucratic control.[111]

---

expressiveness of the Etruscan vases and of the rhinoceros, see, in this book, Camilla Flodin, 'In the Name of the Rhinoceros: Expression beyond Human Intention', pp. 19–42

108 Flodin, 'In the Name of the Rhinoceros: Expression beyond Human Intention', p. 32.

109 Cook, *Adorno on Nature*, p. 159.

110 Adorno, *Ästhetische Theorie*, p. 107; *Aesthetic Theory*, p. 68.

111 On Adorno's idea that works of art express the voice of the victims of the domination of external and internal nature, see Gerhard Schweppenhäuser,

4. *Humanism and naturalism in Adorno: a retrospective starting from* Negative Dialectics

According to Steven Vogel, in Adorno's late work – such as in *Dialectic of Enlightenment* – nature continues to appear as a problem. It has to be recognised as an 'externality'; nevertheless, access to it is foreclosed to humans because the external world with which they are in contact is always the product of their mediating acts.[112] In a passage at the end of *Dialectic of Enlightenment*, Horkheimer and Adorno affirm that '[n]ature in itself is neither good, as was believed by the old Romanticism, nor noble, as is asserted by the new'.[113] They add: 'What threatens the prevailing praxis and its inescapable alternatives is not nature, with which that praxis coincides, but the remembrance of nature'.[114] The fact that nature has to be only 'remembered' means that nature could stimulate social critique 'only as a conceptual memory of the nonconceptual'; it cannot constitute an effective and understandable model in order to construct a social order.[115] Indeed, in his later work, Adorno attempts to conceive of nature, not in a fetishist and nostalgic way, typical of a certain romanticism, nor as a superior, purer and more genuine reality, to be contrasted with the artificiality of society.[116] He reflects on the nonconceptual while still using concepts and in

'"Jenseits der ästhetischen Immanenz". Natur und Gesellschaft in Adornos Theorie des Ästhetischen', *Scenari*, 14 (2021), 239–58.

112 Vogel, *Against Nature*, p. 84.

113 Horkheimer and Adorno, *Dialektik der Aufklärung*, p. 287; *Dialectic of Enlightenment*, pp. 211–12.

114 Horkheimer and Adorno, *Dialektik der Aufklärung*, p. 287; *Dialectic of Enlightenment*, p. 212.

115 Vogel, *Against Nature*, p. 85.

116 See Adorno, *Negative Dialektik*, p. 158; *Negative Dialectics*, pp. 155–6: 'Karl Kraus's line "The origin is the goal" sounds conservative, but it also expresses something that was scarcely meant when the line was uttered: namely, that the concept "origin" ought to be stripped of its static mischief. Understood this way, the line does not mean that the goal had better make its way back to the origin, to the phantasm of "good" nature; it means that nothing is original except the goal, that it is only from the goal that the origin will constitute itself. There is no origin save in ephemeral life'.

the awareness that an immediate nature, independent of human beings who manipulate and transform the material world, does not exist. The attention paid to the nonconceptual and to the subject as an element of nature, is what makes a different and non-dominant relationship between human beings and the natural environment possible. It is essential that in this attempt to grasp nature beyond the domination relationship, nature appears as 'what is not made'.[117] This expression suggests the idea of a relationship to nature where the subject is capable of perceiving something not identical to it but not completely recognised or identified.[118]

In *Negative Dialectics*, the question of nature must be addressed within the framework of the 'non-identity'. The thematisation of the 'non-identical' continues, with further theoretical tools, the work of *Dialectic of Enlightenment* and other texts previously examined, with a view to correcting anthropocentric humanism. As Eric Nelson points out, in *Negative Dialectics*, Adorno argues 'for the interruptive and critical significance' of 'non-identity' and 'nature'. 'Non-identity' refers to alterity, difference, and otherness: the object or subject-matter irreducible to, and potentially fracturing, thought. 'Nature' designates animality, materiality, sensuousness, and the worldly basis of human praxis and reflection.[119] The non-identity of concept and object could be considered as the pivot of Adorno's entire philosophical enterprise:[120] 'The name of dialectics says no more, to begin with, than that objects do not go into their concepts without leaving a remainder, that they come to contradict the traditional norm of adequacy. Contradiction [...] indicates the untruth of identity [*ist Index der Unwahrheit von Identität*], the fact

---

117 Adorno, *Ästhetische Theorie*, p. 98; *Aesthetic Theory*, p. 62.

118 See Morgan, 'Reconciliation with Nature: Adorno on Reason, Nature and Critique', p. 28. According to Alastair Morgan, the idea of a reconciliation with nature is plausible to the extent that a unilateral conception of self-preservation is overcome: self-preservation should not be implemented as a struggle against nature but should be based on a rationale that, far from seeking to dominate the natural environment, works and cooperates with the forces of nature. Ibid., p. 30.

119 Nelson, 'Revisiting the Dialectic of Environment', p. 106.

120 Cook, *Adorno on Nature*, p. 13.

that the concept does not exhaust the thing conceived [*des Aufgehens des Begriffenen in Begriff*]'.[121]

As has been shown above (§ 2 of this chapter), Adorno criticises the radical separation of the subject from the object, which implies that 'subject reduces the object to itself'.[122] Indeed, he puts forward a critical principle: that of reflecting on the tendency of the subject of knowledge to reduce the known object to the concepts by which it is known, and on the violence that identificatory thinking inflicts on this object.[123] He stresses the object's preponderance over the subject: if the object can be conceived only by a subject, it always 'remains something other than the subject'.[124] It means that nature preponderates, although humans persist in thinking of themselves as completely other than, and separate from, nature. This thesis reminds us that the subject is embedded in something else, that can be designated as 'nature' – on the other hand, the ability to reflect on the fact that human beings are part of nature triggers the possibility of not blindly following natural goals (dictated by self-preservation) and of taking alternative actions.[125] Similarly, the 'somatic element', the 'bodily moment', reminds the mind of its anchorage in the body. This is not a 'naïve naturalism', but the concrete problem of suffering. Indeed, the slightest trace of senseless suffering in the

121 Adorno, *Negative Dialektik*, pp. 16–17; *Negative Dialectics*, p. 5. On this point, see Cook, *Adorno on Nature*, p. 13.

122 Adorno, *Stichworte. Kritische Modelle 2*, 'Zu Subjekt und Objekt', pp. 741–758 (p. 742); *Critical Models: Interventions and Catchwords*, 'Catchwords: Critical Models 2', 'On Subject and Object', pp. 245–58 (p. 246).

123 See Emmanuel Renault, 'Sujet-objet: le dispositif Hegel-Kant', *Cahiers philosophiques*, 154.3 (2018), 9–28 (pp. 20 and 23).

124 'An object can be conceived only by a subject but always remains something other than the subject, whereas a subject by its very nature is from the outset an object as well. Not even as an idea can we conceive a subject that is not an object; but we can conceive an object that is not a subject. To be an object also is part of the meaning of subjectivity; but it is not equally part of the meaning of objectivity to be a subject'. Adorno, *Negative Dialectics*, p. 184; *Negative Dialectics*, p. 183.

125 Adorno, *Probleme der Moralphilosophie*, p. 154; *Problems of moral philosophy*, pp. 103–4.

society 'belies all the identitarian philosophy that would talk us out of that suffering'.[126] Moreover, the blind and frantic pursuit of individual self-preservation was generated by dismay at the external natural world. In this sense, what is historical is also to be considered natural because, in light of its violent origins, 'it remains under the spell of a blind nature'.[127] The core thesis of *Negative Dialectics* and the arguments on which it is based constitute one of the main points of Adorno's critique of idealism. It is the critique of the fallacy of constitutive subjectivity: it means that 'the mind is not primary'.[128] Indeed, Adorno states that Hegel himself derives self-conscious mind from matter. Hypostatizing the mind, Hegel was nonetheless barely able to conceal the origin of the 'I' in the 'Not-I'. Even for Hegel, mind (*Geist*) ultimately originates 'in the real living process, in the law of the survival of the species, of providing it with nutriments'.[129]

Two clarifications are necessary in this regard. Firstly, as Deborah Cook states, the preponderance of the object does not simply mean that nature preponderates over individuals. Indeed, the objectivity that cannot be reduced to the subjectivity, and affects thoughts and behaviour of individuals, is not merely natural, but social and historical. Arguing the preponderance of the material, natural world in the relation between object and subject, Adorno also emphasizes a specific kind of preponderance or burden imposed by society on individuals.[130] Referring to Marx, he notes that two things are intertwined 'in the realm of things [*im Dinghaften*]': the non-identity

126 Adorno, *Negative Dialektik*, p. 202–3; *Negative Dialectics*, p. 203. On these topics, see Estelle Ferrarese, *The Fragility of Concern for Others: Adorno and the Ethics of Care*, trans. by Steven Corcoran (Edinburgh: Edinburgh University Press, 2020). Among other things, the author highlights the role that the corporeal moment, a strong idea of the body and a materialist morality play in Adorno's works. On the somatic moment as a moment that cannot be reduced to the purely cognitive sphere and on its political and aesthetic implications, see Leonardo Distaso, *Marcuse, Adorno. Percorsi tra estetica e politica* (Rome: Carocci, 2022), pp. 161–171.

127 Adorno, *Zur Lehre von der Geschichte und von der Freiheit*, p. 179; *History and Freedom*, p. 124.

128 Cook, *Adorno on Nature*, p. 9.

129 Adorno, *Negative Dialectics*, p. 199; *Negative Dialectics*, p. 198.

130 Cook, *Adorno on Nature*, p. 13.

of the object and the submission [*die Unterwerfung*] of human beings to prevailing relations of production, to their own functional context, which they are unable to know.[131] Secondly – and in connection with his analysis of Marx's remarks on the character of a liberated society[132] –, as Alfred Schmidt points out, 'Adorno always warned against glorifying "primitive, pre-labour conditions"; rather, it is one of the essential aspects [...] of real humanism to realise the human through the material. All attributes ascribed to Nature herself, rather than to her truly humanized form, are unable to avert the fate that has befallen humanity, but rather lead deeper into it. If nature is suddenly elevated to a role model and goal, as in the national-authoritarian ideology, then "it signifies the spirit of contradiction, lies and bestiality"'.[133] These considerations recall the passage from *Dialectic of Enlightenment*, quoted at the beginning of this paragraph, which warns against a nostalgic, romantic attitude towards nature, which can convey forms of brutality and inhumanity. According to Schmidt's interpretation, Adorno contrasts this nostalgic attitude towards a lost primordial nature with a 'real humanism'.[134] This notion refers to

131 Adorno, *Negative Dialektik*, p. 192; *Negative Dialectics*, p. 192.

132 'The mature Marx, in his few remarks on the character of a liberated society, changed his position on the cause of reification, the division of labor. He now distinguished the state of freedom from original immediacy. In the moment of planning – the result of which, he hoped, would be production for use by the living rather than for profit, and thus, in a sense, a restitution of immediacy – in that planning he preserved the alien thing; in his design for a realization of what philosophy had only thought, at first, he preserved its mediation'. Adorno, *Negative Dialektik*, pp. 192–3; *Negative Dialectics*, p. 192.

133 Alfred Schmidt, *Kritische Theorie, Humanismus, Aufklärung. Philosophische Arbeiten 1969-1979* (Stuttgart: Philipp Reclam Jun., 1981), p. 48 (translation by the author). Here Schmidt quotes from Theodor W. Adorno, *Dissonanzen* (Göttingen: Vandenhoeck & Ruprech, 1956), p. 57.

134 According to Schmidt, Adorno's concept of 'real humanism' refers to an 'idea of real humanity, which, in contrast to traditional bourgeois humanity, is not only rooted in a world of beautiful appearances removed from the pressures of everyday life, but is embodied in tenacious historical work aimed at bringing about a state that is finally beyond the conflict between material practice and the cultural sphere. To the extent, of course,

an understanding and orientation towards humanity that cannot be precisely defined through the philosophies of humanism and anti-humanism as opposing perspectives.[135] The correct attitude towards nature is not one that seeks an original, uncontaminated nature, but rather one based on the awareness of the intertwining of human activity (historical) and nature in its historical development.[136] In the view of Adorno, who here seems to echo Marx's *Economic and Philosophic Manuscripts of 1844*,[137] such an interconnection lies in the fact that '[l]abor – and in the last analysis its reflective form, spirit, as well – cannot be conceived without the concept of nature, any more than can nature without labor: the two are distinct from

that people in this sphere, especially artists, succeed in truly experiencing a moment of social helplessness linked to the disenchantment of culture with the immediate world of haggling, they simultaneously succeed in transforming that detachment into critical distance, powerlessness into strength. Therefore, the programmatic idea of an "art of real humanism" appears repeatedly in Adorno's writings, especially in his early musical writings'. Schmidt, *Kritische Theorie, Humanismus, Aufklärung*, p. 31. See Theodor W. Adorno, *Moments musicaux: neu gedruckte Aufsätze 1928-1962* (Frankfurt am Main: Suhrkamp, 1964), p. 160; *GS* 17: *Musikalische Schriften* IV, *Moments musicaux. Impromptus* (Frankfurt am Main: Suhrkamp, 1982), p. 139.

135 Alice Nilsson, 'The humanism of critical theory: The Frankfurt School's "realer humanismus"', *Philosophy & Social Criticism*, 51.5 (2023), 727–46 (p. 728).

136 Similarly, as Adorno states in a 1959 text on German sociology, science should understand interpersonal relationships by emphasising their function in the 'metabolic interchange with nature and with social totality'. Theodor W. Adorno, 'Zum gegenwärtigen Stand der deutschen Soziologie', in Adorno, *GS* 8, *Soziologische Schriften* I, 4th edition (Frankfurt am Main: Suhrkamp, 2018), pp. 500–31 (p. 504). Translation by the author. This text also shows that '"[n]ature" is always a relational term for Adorno'. Gillian Rose, *The Melancholy Science: An Introduction to the Thought of Theodor W. Adorno* (London: Verso, 2014), p. 102.

137 'The worker can create nothing without nature, without the *sensuous external world*. It is the material on which his labor is manifested, in which it is active, from which and by means of which it produces'. Karl Marx, *Economic and Philosophic Manuscripts of 1844*, trans. by Martin Milligan (New York: Prometheus Book, 1988), p. 72.

and mediated by one another at the same time'.[138] As Marx writes in his *Manuscripts*, the authentic (i.e., neither romantic nor nostalgic) resolution of the conflict between human beings and nature and between man and man, lies in this interconnection – and, notably, in the awareness of it. From this perspective, it is communism, as complete naturalism, that reveals itself as true humanism.[139] In other words, on the Marx-Adorno axis, the possibility of a humanism that is neither idealistic (because it is based on the material conditions of human existence and on the preponderance of objects and bodies) nor anthropocentric, becomes intelligible: it considers human beings in their inescapable, dynamic, and plastic relationship with the material, natural world.

Indeed, just as in the article 'Kultur and Culture' Adorno envisages the possibility of a relationship between human beings and nature based on 'care' and 'caring', rather than domination and exploitation, in *Negative Dialectics* he suggests the idea of a way of thinking that seeks to go beyond the 'given', which aims to compensate for what it has violated by trying to trap it in its logical-conceptual grids:

> Thought forms tend beyond that which merely exists, is merely 'given'. The point which thinking aims at its material is not solely a spiritualized control of nature. While doing violence to the object of its syntheses, our thinking heeds a potential that waits in the object, and it unconsciously obeys the idea of making amends to the pieces for

---

138 Theodor W. Adorno, *Hegel: Three Studies*, trans. by Shierry Weber Nicholsen (Cambridge, MA: MIT Press, 1993), p. 23.

139 '*Communism* as the *positive* transcendence of *private property* as *human self-estrangement,* and therefore as the real *appropriation of the human* essence by and for man; communism therefore as the complete return of man to himself as a *social* (i.e., human) being – a return accomplished consciously and embracing the entire wealth of previous development. This communism, as fully developed naturalism, equals humanism, and as fully developed humanism equals naturalism; it is the *genuine* resolution of the conflict between man and nature and between man and man – the true resolution of the strife between existence and essence, between objectification and self-confirmation, between freedom and necessity, between the individual and the species. Communism is the riddle of history solved, and it knows itself to be this solution'. Marx, *Economic and Philosophic Manuscripts of 1844*, p. 102–3.

> what it has done. In philosophy, this unconscious tendency becomes conscious.[140]

If the telos of philosophy (i.e., the open and the unguaranteed) is unsystematic,[141] philosophy should be able to recognise the intolerance towards that which is other than itself, an intolerance that has punctuated its history, especially in the modern age:

> Great philosophy was accompanied by a paranoid zeal to tolerate nothing else, and to pursue everything else with all the cunning of reason, while the other kept retreating farther and farther from the pursuit. The slightest remnant of nonidentity sufficed to deny an identity conceived as total. The excrescences of the systems, ever since the Cartesian pineal gland and the axioms and definitions of Spinoza, already crammed with the entire rationalism he would then deductively extract – by their untruth, these excrescences show the untruth, the mania, of the systems themselves.[142]

Philosophy that aims to remain faithful to its most authentic vocation must denounce the totalitarian nature of systems. The system is nothing more than the expository structure of a totality that encompasses everything and to which nothing remains external. In this way, it is idealistic by nature; it absolutizes the thought with respect to its contents and sublimates the content into thought itself, thereby denying it in its otherness to thought.[143] As Adorno writes in a subsequent paragraph of *Negative Dialectics*, '[w]hat tolerates nothing that is not like itself thwarts the reconcilement for which it mistakes itself. The violence of equality-mongering reproduces the contradiction it eliminates'.[144] Negative dialectics reveals the split, 'the untruth' of the model of totality constructed by idealism: it aims to achieve compensation, reparation. 'The subject must make up for [*wiedergutmachen*] what it has done to nonidentity [*am Nichtidentischen*]. This is precisely what liberates it from

140 Adorno, *Negative Dialektik*, pp. 30–1; *Negative Dialectics*, p. 19.
141 Adorno, *Negative Dialektik*, p. 31; *Negative Dialectics*, p. 20.
142 Adorno, *Negative Dialektik*, p. 33; *Negative Dialectics*, p. 22.
143 Adorno, *Negative Dialektik*, p. 35; *Negative Dialectics*, p. 24.
144 Adorno, *Negative Dialektik*, p. 146; *Negative Dialectics*, pp. 142–3.

the semblance of its absolute being-for-itself [*seines absoluten Fürsichseins*]'.[145] In order to understand what identity has done to the non-identity, it is necessary to focus on the concept of causality. According to Adorno, causality is 'the spell of dominated nature [*der Bann der beherrschten Natur*]'; it is based on identity which, as a principle established by the subject, constitutes the spiritual reflection of the real domination of nature.[146] When the subject imposes the concept of causality upon a material content, it is driven by a compelling urge to identify objects with its causal conceptions of them, by imposing legality and regularity upon them: 'Causality hypostatizes the form, as binding upon a content which on its own would not assume that form'.[147] In reality, the causal nature's order, the legality imposed on it, is the conceptual form of the way in which subject has appropriated nature and understood it in certain instrumentally useful terms.[148]

Nevertheless, the ability of negative dialectics to unveil the mechanism of domination of nature on the level of legality imposed by the categories of subjectivity, does not mean that it is purely and simply the way out, the solution to the model of thought that mutilates nature in an attempt to yoke it to its conceptual structures. It certainly opposes totality, unmasking its non-identity with itself, and opposes the identifying principle that has its ancestor in the principle of exchange, in the reduction of human labour to an abstract element measured by average working time.[149] Yet, precisely in order to oppose totality and identifying thought, negative dialectics remains tied to the highest categories of the philosophy of identity. From this perspective, it too is false, as it is of an identitarian-logical nature, and must therefore be corrected in its critical approach.[150] After all, as Adorno writes in another key paragraph of the section

145 Adorno, *Negative Dialektik*, p. 149; *Negative Dialectics*, pp. 145–6.

146 Adorno, *Negative Dialektik*, p. 266; *Negative Dialectics*, p. 269.

147 Adorno, *Negative Dialektik*, p. 232; *Negative Dialectics*, p. 234.

148 See Craig Reeves, 'Causality and Critical Theory: Nature's Order in Adorno, Cartwright and Bhaskar', *Journal of Critical Realism*, 8.3 (2009), 316–342 (p. 325).

149 Adorno, *Negative Dialektik*, pp. 149–50; *Negative Dialectics*, pp. 146–7.

150 Adorno, *Negative Dialektik*, pp. 150; *Negative Dialectics*, p. 147.

'Negative Dialectic. Concepts and Categories', entitled 'Cogitative self-reflection', one cannot think without identification, since thinking is determining, and every determination is identification. Nevertheless, in reality, the secret *telos* of identification is non-identity (which is, therefore, what is to be saved in it), so traditional thinking errs insofar as it considers identity to be its purpose. Identitarian thinking reduces the thing to an example of a class, expressing 'what something comes under, what it exemplifies or represents, and what, accordingly, it is not itself'. While cognition of non-identity is dialectic: 'it identifies to a greater extent, and in other ways, than identitarian thinking'; it aims 'to say what something is'.[151] The claim that the thing-in-itself corresponds to the concept is a form of '*hubris*', just as the simple identifying judgement contains a 'pragmatist, nature-controlling element [*pragmatistischen, naturbeherrschenden Element*]'. Conversely, the reproach that the thing is not identical to the concept reveals the desire that it could become so, that the antagonism between thing and concept could be overcome. The awareness of non-identity then also detects a utopian element in the identifying judgement: 'Utopia would be above identity and above contradiction; it would be a togetherness of diversity [*ein Miteinander des Verschiedenen*]'.[152]

151 Adorno, *Negative Dialektik*, pp. 152; *Negative Dialectics*, p. 149.

152 Adorno, *Negative Dialektik*, pp. 152–3; *Negative Dialectics*, pp. 149–50. The contrast, the antagonism between concept and object then takes on its own specific meaning in practical philosophy. In his course *Problems in Moral Philosophy*, Adorno examined how Kant addresses the dualism of mind and nature from a moral point of view. The spirit – in Kant's argument – is not simply part of nature, where what we call 'nature' is what is constituted in opposition to the experience of our mental states. Adorno, *Probleme der Moralphilosophie*, pp. 152–3; *Problems of moral philosophy*, p. 102. With his theory of freedom in the midst of nature, Kant emphasised what in human beings, in their nature, goes beyond the mere natural element, namely the 'consciousness of self [*der Selbstbesinnung*]'. Indeed, Adorno comments, 'we are no longer simply a piece of nature from the moment we recognize that we are a piece of nature'. Adorno, *Probleme der Moralphilosophie*, p. 154; *Problems of moral philosophy*, p. 103. Nevertheless, Kant does not overcome the dualism of spirit and nature: all categories of ethics are nothing more than 'categories by which the domination of nature is achieved'. Adorno,

Dialectics, which undermines the imposition of identity by means of the energy accumulated within it, coagulated in its objectifications, finds its empirical content in the resistance of the other against identity ['am Widerstand des Anderen Gegen die Identität'].[153] Negative dialectics shows that 'what is, is more than it is', i.e, the entity identified by thought, where this excess is immanent to the entity as that which has been removed from it in order to subsume it under a category: 'In that sense, the nonidentical would be the thing's own identity against its identifications'.[154] In other words, negative dialectics aims to break out of the magic circle of identification without opposing idealism with a mere realist thesis. Such a circle of identification is traced by thinking that tolerates nothing outside itself; it is the product of a totalitarian rationality that, historically, arose as a reaction to threatening nature (*Bedrohlichen der Natur*).[155] Once again, as in *Dialectic of Enlightenment*, Adorno emphasises the link between totalitarian rationality and the original perception of nature as a threat to survival, a link in which reason finds its limit and from which it reveals itself as pathological: 'In fear, bondage to nature [*in der Angst Naturverfallenheit*] is perpetuated by a thinking that identifies, that equalizes everything unequal. Thoughtless rationality is blinded to the point of madness by the

---

Probleme der Moralphilosophie, p. 155; *Problems of moral philosophy*, p. 104. In *Negative Dialectics*, Adorno shows how being part of nature and perceiving one's powerlessness in the face of society constitute the litmus test of the limits of human freedom: 'What makes the subjects aware of the bounds of their freedom is that they are part of nature, and finally, that they are powerless against society, which has become independent of them. [...] The subjects are not only fused with their own physical nature; a consistent legality holds sway also in the psychological realm, which reflection has laboriously divided from the world of bodies'. Adorno, *Negative Dialektik*, p. 220; *Negative Dialectics*, p. 221. The mirage of freedom – 'which reason will not allow to be withered by any proof of causal interdependence – is the phantasm of reconciling nature and the mind'. Adorno, *Negative Dialektik*, p. 228; *Negative Dialectics*, p. 229.

153 See Adorno, *Negative Dialektik*, pp. 159 and 163; *Negative Dialectics*, pp. 157 and 160–1.

154 Adorno, *Negative Dialektik*, p. 164; *Negative Dialectics*, p. 161.

155 Adorno, *Negative Dialektik*, p. 174; *Negative Dialectics*, p. 172.

sight of whatsoever will elude its rule. For the present, reason is pathic; nothing but to cure ourselves of it would be rational'.[156] Indeed, the subjugation of nature for human purposes is itself a natural relationship – connected, as has been shown above, with the impulse of self-preservation. Therefore, the supposed supremacy of reason dominating nature is only apparent, and the subject who believes he is exercising dominion becomes part of what he intends to dominate.[157]

Adorno develops this further in an illuminating paragraph in the third part of *Negative Dialectics*, entitled 'Kant's intelligible character'. Here he highlights the dialectical character of the connection between reason and nature:

> The prehistory of reason, that it is a moment of nature and yet something else, has become the immanent definition of reason. It is natural as the psychological force split off for purposes of self-preservation; once split off and contrasted with nature, it also becomes nature's otherness. But if that dialectic irrepressibly turns reason into the absolute antithesis of nature, if the nature in reason itself is forgotten, reason will be self-preservation running wild and will regress to nature. It is only as reflection upon that self-preservation that reason would be above nature.[158]

This regression in nature is not surprising, since self-preservation, Spinoza's *sese conservare*, is a law of nature for every living being. Ultimately, it reveals the tautology of identity: what ought to be is what already is; the will can only desire the continuation of the will of the willing, which, as a pure means of itself, becomes the end of the will.[159] If human action ultimately refers back to biology, to the tendency to preserve oneself and to desire nothing other than such preservation, human history, which is marked by the progressive control of nature, 'continues the unconscious history of nature, of devouring and being devoured'.[160] Once again, as has already been

156 Adorno, *Negative Dialektik*, p. 174; *Negative Dialectics*, p. 172.
157 Adorno, *Negative Dialektik*, p. 181; *Negative Dialectics*, pp. 179–80.
158 Adorno, *Negative Dialektik*, p. 285; *Negative Dialectics*, p. 289.
159 Adorno, *Negative Dialektik*, p. 342; *Negative Dialectics*, p. 349.
160 Adorno, *Negative Dialektik*, pp. 348–9; *Negative Dialectics*, p. 355.

shown in the first chapter of this book, Adorno evokes Marx's thesis of the mutual conditioning of the history of nature and the history of mankind,[161] emphasising the ambivalence of the antithesis between nature and history. This antagonism proves to be both true and false: 'true insofar as it expresses what happened to the natural element; false insofar as, by means of conceptual reconstruction, it apologetically repeats the concealment of history's natural growth by history itself'.[162] Once again, in these pages of *Negative dialectics*, Adorno advocates a natural history, a concept that shows the dialectical relation of nature to history.[163] Indeed, '[s]uch a social concept of nature has a dialectic of its own'.[164] Here, it is possible to detect the theme of 'reconciliation with nature', which remains not only the 'guiding idea of Adorno's philosophy of history'[165] but also a core element of his entire body of work. However unclear Adorno might have been about the elements of this reconciliation, the 'mindfulness [*Eingedenken*] and self-reflection [...] which breaks the immediate power over nature'[166] as well as the idea that '*Humanität* is mindfulness [*das Eingedenken*] of nature in human beings'[167] express the need to undermine and overcome a model of reason 'which reifies relations of nature and society'.[168]

---

161 'We know only a single science, the science of history. One can look at history from two sides and divide it into the history of nature and the history of men. The two sides are, however, inseparable; the history of nature and the history of men are dependent on each other so long as men exist'. Marx and Engels, *Die deutsche Ideologie*, p. 18; *The German Ideology*, p. 28.

162 Adorno, *Negative Dialektik*, p. 351; *Negative Dialectics*, p. 358.

163 See Omar Dahbour, 'Totality, Reason, Dialectics: The Importance of Hegel for Critical Theory from Lukács to Honneth', in *The Palgrave Handbook of Critical Theory*, ed. by Michael J. Thompson (New York: Palgrave Macmillan, 2017), pp. 87–108 (p. 106).

164 Adorno, *Negative Dialektik*, p. 349; *Negative Dialectics*, p. 355.

165 Axel Honneth, *The Fragmented World of the Social: Essays in Social and Political Philosophy* (Albany: State University of New York Press, 2005), p. 101.

166 Adorno, 'Kultur und Culture', p. 157; 'Kultur and Culture', pp. 146–7.

167 Adorno, *Noten zur Literatur*, 'Balzac-Lektüre' pp. 139–57 (p. 143); *Notes on literature*, 'Reading Balzac', pp. 132–46 (p. 135).

168 Honneth, *The Fragmented World of the Social*, p. 100.

This project is based on the notion of non-identity, specifically on the recognition of nature's independent reality which cannot be reduced to its manipulation.[169]

169 Andrew Feenberg, 'Why Students of the Frankfurt School Will Have to Read Lukács', in *The Palgrave Handbook of Critical Theory*, pp. 109–33 (p. 120).

CONCLUSIONS

The fundamental position of Adorno and Horkheimer, in relation to the themes explored in our book, as formulated in their most significant texts of the 1940s and also considering subsequent developments up to the end of the 1960s, is based on the refutation of the enlightenment thesis, of Baconian origin, that is: the only real means of liberating humanity from the threat it perceives in nature is technical-scientific control over the environment-world. Adorno and Horkheimer refute this thesis by demonstrating that the achievement of this project of domination did not lead to any form of emancipation, but rather resulted in the worst forms of barbarism. Therefore, if we want to save the enlightenment and its potential for emancipation,[1] we need to transform the concept of reason, which has crystallised as pure instrumental reason, and open ourselves up to criticism of domination and to the idea of reconciling human beings with each other and with nature.[2]

Like Marx, Horkheimer and Adorno define capitalism as a mode of production, but at the cost of profoundly transforming this concept. For Marx, the mode of production was essentially centred on the question of the exploitation of labour. For Horkheimer and Adorno, this is only one aspect of capitalism.[3] Indeed, labour exploitation is just one form among many of the exploitation of nature. It is the economic domination of a working body, to which other forms

1 See the text already mentioned and analysed in the third chapter of this book: Horkheimer and Adorno, ['Rettung der Aufklärung. Diskussionen über eine geplante Schrift zur Dialektik'], pp. 594–605.

2 See Stefano Petrucciani, *Theodor W. Adorno's Philosophy, Society, and Aesthetics* (Cham: Palgrave Macmillan, 2021), p. 25.

3 See Vuillerod, *Theodor W. Adorno. La domination de la nature*, p. 98.

of domination from which profit can be derived should be added: resource extraction, destruction of ecosystems, industrial slaughter of animals, but also appropriation of living beings of all kinds.[4]

Adorno and Horkheimer do not reproduce an anthropocentric discourse that divides the human from the natural material world, which remains – as in Habermas' approach which reiterates the metaphysical division of nature (*Natur*) and spirit (*Geist*) – misunderstood or unrecognized as something other than, and in excess of, human communicative rationality and, therefore, legitimates the misuse and destruction of animals and environments.[5] Eric S. Nelson has suggested that their vision of human and non-human nature offers schemes 'for engaging the contemporary environmental crisis by correcting the anthropocentric humanism, intersubjective constructivism, and deontological idealism of contemporary discourse ethics, dialogical ethics, and social contract theory'.[6]

In the wake of Marx, who criticises Feuerbach's myth of 'pure nature',[7] Horkheimer and Adorno do not represent 'nature' as a utopic or primordial kingdom that human beings have lost and should restore. It is arduous to think of nature as completely separated from social reality, as a separate world, accessible in its purity. What we call nature is always inextricably linked to (and mediated through) the human and social sphere. But, if we critically reflect on naive representation of nature and on the type of relation with non-human entities that Western cultures have realised (that is the domination over nature, linked to the non-true preponderance of the subject over the object), we can start to conceive and realise a different relation to nature, one that reflects particularly on the suffering running through nature and that unites human beings and the other living beings.

Horkheimer and Adorno emphasise man's domination over nature or man's care for the environment. However, they do not seem to consider the relationship between man and nature in a broader

4 Vuillerod, *Theodor W. Adorno. La domination de la nature*, p. 99.
5 Nelson, 'Revisiting the Dialectic of Environment', p. 106.
6 Nelson, 'Revisiting the Dialectic of Environment', p. 106.
7 See Henri Lefebvre, *Dialectical Materialism*, trans. by John Sturrock (London: Jonathan Cape, 1968), pp. 67–8.

context, such as that outlined by Lovelock and Latour:[8] the idea of a plurality of agents on Earth, an intentionality extended to all agents, the idea of unintentional feedback. In other words, it is a question of exploring how natural elements react to 'human domination', highlithing the reactions (whether intentional or not) and the interests of non-human organisms. Therefore, from a certain point of view, one could argue that Horkheimer and Adorno's criticism of an anthropocentric view of the world and of the relationship of domination that exists between humans and nature, is itself human–centred: they overlook the agency of non-human beings, their ability to react to human actions, that is, as Latour stresses, the 'power of invention and surprise' of 'the nonhumans'.[9]

However, the nexus between 'humanism' and (the reflection on) 'nature' in Horkheimer and Adorno's works emerges to the extent that, also with a view to countering human–centred theoretical perspectives, they seek to emphasise the rational potential of the principle of self-preservation, which consists in its ability to produce a reflective return of nature upon itself. This always happens in a dialectical process: indeed, if, on the one hand – as Adorno writes in 'Zum Verhältnis von Soziologie und Psychologie' (1955) – the Self is also instinct,[10] on the other hand, all human instincts, as with everything else that exists, do not remain unchanged: they are historically conditioned and, therefore, end up being profoundly

---

8 Latour, *Face à Gaïa*, pp. 128–9; *Facing Gaia*, pp. 97–8. 'No agent on Earth is simply superimposed on another like a brick juxtaposed to another brick. On a dead planet, the components would be placed partes extra partes; not on Earth. Each agency modifies its neighbors, however slightly, so as to make its own survival slightly less improbable. This is where the difference between geochemistry and geobiology lies. It means not that Gaia possesses some sort of 'great sensitive soul,' but that the concept of Gaia captures the distributed intentionality of all the agents, each of which modifies its surroundings for its own purposes'. Latour, *Face à Gaïa*, p. 129; *Facing Gaia*, p. 98.

9 Latour, *Face à Gaïa*, p. 139; *Facing Gaia*, p. 108.

10 'The self, as something that has arisen, is a piece of instinct and at the same time something else'. Theodor W. Adorno, 'Zum Verhältnis von Soziologie und Psychologie', in Adorno, *GS* 8, pp. 42–85 (p. 84). Translation by the author.

transformed in the historical process.[11] As seen in the last paragraph of Chapter 4, reason itself 'is a moment of nature and yet something else': it splits off from nature for the purpose of self-preservation, becoming, in this way, 'nature's otherness'. 'But – Adorno explains – if that dialectic irrepressibly turns reason into the absolute antithesis of nature, if the nature in reason itself is forgotten, reason will be self-preservation running wild and will regress to nature. It is only as reflection upon that self-preservation that reason would be above nature'.[12] If nature should not be forgotten in reason, symmetrically – as Adorno writes in 'Reading Balzac' – humanity maintains a special bond with nature: '*Humanität* is mindfulness of nature [*das Eingedenken der Natur*] in human beings'.[13] The remembrance of nature should not be understood as a return to an original naturalness in contrast to the humanistic reification of man,[14] but as a revitalisation of nature within the social and historical relationship that connects it to the subject.[15] Considering that, as mentioned above, the human is not yet, for Adorno, something definable and known, and that 'the human being is a result, not an εἶδος',[16] *Eingedenken* should be

11 See Adorno, *Zur Lehre von der Geschichte und von der Freiheit*, p. 328; *History and Freedom*, p. 236.

12 Adorno, *Negative Dialektik*, p. 285; *Negative Dialectics*, p. 289.

13 Adorno, *Noten zur Literatur*, 'Balzac-Lektüre' p. 143; *Notes on literature*, 'Reading Balzac', p. 135.

14 This seems to be Martin Desrosiers' hypothesis. See Desrosiers, 'Un humanisme négatif. Horkheimer, Adorno et la critique du "kitsch humaniste"', p. 304.

15 According to Gilles Moutot, insofar as it is conceived in terms of remembrance, and therefore of its intertwining with history, the subject's experience of nature appears to be linked to a critical and normative aim that nevertheless refrains from 'any naturalistic overload [*toute surcharge naturaliste*]'. See Gilles Moutot, '"Le souvenir de la nature dans le sujet". Une actualité de La Dialectique de la raison', *Astérion*, 7 (2010), https://journals.openedition.org/asterion/1622.

16 Adorno, *Stichworte. Kritische Modelle 2*, 'Zu Subjekt und Objekt', p. 758; *Critical Models: Interventions and Catchwords*, 'Catchwords: Critical Models 2', 'On Subject and Object', p. 258. See also Adorno, *Probleme der Moralphilosophie*, pp. 250–1; *Problems of moral philosophy*, p. 169. On this point, see Deborah Cook, *Adorno, Foucault and the Critique of the West* (London and New York, 2018), pp. 113–14.

interpreted in a dynamic key and considered in the socio-historical process – as Horkheimer also emphasises in 'Traditional and Critical Theory'[17] – in which human beings and natural entities always interact and mediate with each other. Put another way, 'nature itself has a historical element, in the dual sense that man conceives of it differently at different times and that he actively works to change it'.[18] Finally, revealing itself to be a reflective recognition by human beings of the natural element within them, *Eingedenken* goes hand in hand with the humble recognition of dominion over nature, which – as has been shown in the third chapter of this book – is advocated in *Dialectic of Enlightenment*.[19]

617: See also James Gordon Finlayson, 'Adorno's Metaphysics of Moral Solidarity in the Moment of its Fall', in *A Companion to Adorno*, ed. by Peter E. Gordon, Espen Hammer and Max Pensky (Hoboken, NJ: Wiley Blackwell, 2020), pp. 615–30 (p. 617): 'human nature is not something fixed and immutable that unfolds wholly under its own steam. For human beings, through their labor, can affect, and to an extent create, the conditions under which their nature is formed and reformed, and thus influence the process by which they become the humanity that they potentially are'.

17 'The facts which our senses present to us are socially preformed in two ways: through the historical character of the object perceived and through the historical character of the perceiving organ. Both are not simply natural; they are shaped by human activity, and yet the individual perceives himself as receptive and passive in the act of perception'. Horkheimer, 'Traditionelle und kritische Theorie', p. 174; 'Traditional and Critical Theory', p. 200.

18 Martin Jay, *The Dialectical Imagination: A History of the Frankfurt School and the Institute of Social Research, 1923-1950* (Boston, MA: Little, Brown and Co., 1973), pp. 53–4.

19 'In the mastery of nature, without which mind does not exist, enslavement to nature persists. By modestly confessing itself to be power and thus being taken back into nature, mind rids itself of the very claim to mastery which had enslaved it to nature'. Horkheimer and Adorno, *Dialektik der Aufklärung*, p. 63; *Dialectic of Enlightenment*, p. 31.

BIBLIOGRAPHY

Primary literature

Max Horkheimer's collected works

Gesammelte Schriften, ed. by Alfred Schmidt and Gunzelin Schmid Noerr (Frankfurt am Main: Fischer, 1985–1996).

Vol. 1: *Aus der Pubertät. Novellen und Tagebuchblätter. 1914-1918* (Frankfurt am Main: Fischer, 1988).
Vol. 2: *Philosophische Frühschriften. 1922-1932* (Frankfurt am Main: Fischer, 1987).
Vol. 3: *Schriften. 1931-1936*, 2nd edition (Frankfurt am Main: Fischer, 1988, 2009).
Vol. 4: *Schriften 1936-1941*, 2nd edition (Frankfurt am Main: Fischer, 1988, 2009).
Vol. 5: *Dialektik der Aufklärung und Schriften 1940-1950*, 2nd edition (Frankfurt am Main: Fischer, 1987, 2014).
Vol. 6: *Zur Kritik der instrumentellen Vernunft und Notizen 1949-1969*, 2nd edition (Frankfurt am Main: Fischer, 1991, 2008).
Vol. 7: *Vorträge und Aufzeichnungen. 1949-1973: 1. Philosophisches; 2. Würdigungen; 3. Gespräche*, 2nd edition (Frankfurt am Main: Fischer, 1985, 2014).
Vol. 8: *Vorträge und Aufzeichnungen 1949-1973: 4. Soziologisches; 5. Universität und Studium* (Frankfurt am Main: Fischer, 1985).
Vol. 9: *Nachgelassene Schriften 1914-1931*: *1. Vorlesung über die Geschichte der neueren Philosophie* (Frankfurt am Main: Fischer, 1987).
Vol. 10: *Nachgelassene Schriften 1914-1931: 2. Vorlesung über die Geschichte der deutschen idealistischen Philosophie [Frankfurt, Wintersemester 1925/26]; 3. Einführung in die Philosophie der*

Gegenwart [Frankfurt, Sommersemester 1926]; Zur Emanzipation der Philosophie von der Wissenschaft (Frankfurt am Main: Fischer, 1990).
Vol. 11: *Nachgelassene Schriften 1914-1931: 4. Aufzeichnungen und Vorträge; 5. Notizen; 6. Poetische Versuche; 7. Diskussionsprotokoll* (Frankfurt am Main: Fischer, 1987).
Vol. 12: *Nachgelassene Schriften 1931-1949*: *1. Vorträge und Aufsätze; 2. Memoranden*; *3. Aufzeichnungen und Entwürfe; 4. Poetische Versuche; 5. Diskussionsprotokolle* (Frankfurt am Main: Fischer, 1985).
Vol. 13: *Nachgelassene Schriften 1949-1972: 1. Vorträge und Ansprachen; 2. Gespräche; 3. Würdigungen; 4. Vorlesungsnachschriften* (Frankfurt am Main, Fischer, 1989).
Vol. 14: *Nachgelassene Schriften 1949-1972: 5. Notizen [Nachgelassene Notizen 1949–1969; Verstreute Aufzeichnungen 1950-1971; Späne. Notizen über Gespräche mit Max Horkheimer, in unverbindlicher Formulierung aufgeschrieben von Friedrich Pollock]* (Frankfurt am Main: Fischer, 1988).
Vol. 15: *Briefwechsel 1913-1936* (Frankfurt am Main: Fischer, 1995).
Vol. 16: *Briefwechsel 1937-1940* (Frankfurt am Main: Fischer, 1995).
Vol. 17: *Briefwechsel 1941-1948* (Frankfurt am Main: Fischer, 1996).
Vol. 18: *Briefwechsel 1949-1973* (Frankfurt am Main: Fischer, 1996).
Vol. 19: *Nachträge, Verzeichnisse und Register*, ed. by Gunzelin Schmid Noerr with the collaboration of Jan Baars, Rainer Clodius, Frederik van Gelder (Frankfurt am Main: Fischer, 1996).

Max Horkheimer's works in English

'The End of Reason', *Studies in Philosophy and Social Science*, 9:3 (1941), 366–88.
Eclipse of Reason (New York: Oxford University Press, 1947).
'The Concept of Man', in Max Horkheimer, *Critique of Instrumental Reason: Lectures and Essays Since the End of World War II*, trans. by Matthew J. O'Connell and others (New York: The Seabury Press, 1974), pp. 1–33.
Dawn and Decline: Notes 1926-1931 and 1950-1969, trans. by Michael Shaw with an afterword by Eike Gebhardt (New York: The Seabury Press, 1978).
'Egoism and Freedom Movement: On the Anthropology of the Bourgeois Era', in Horkheimer, *Between Philosophy and Social Science: Selected Early Writings*, trans. by G. Frederick Hunter, Matthew S. Kramer, and John Torpey (Cambridge, MA: The MIT Press, 1993), pp. 49–110.

'Materialism and Morality', in Max Horkheimer, *Between Philosophy and Social Science: Selected Early Writings* (Cambridge, MA: The MIT Press, 1993), pp. 15–47.

'Montaigne and the Function of Skepticism', in Horkheimer, *Between Philosophy and Social Science: Selected Early Writings* (Cambridge, MA: The MIT Press, 1993), pp. 265–311.

'On the Problem of Truth', in Horkheimer, *Between Philosophy and Social Science: Selected Early Writings* (Cambridge, MA: The MIT Press, 1993), pp. 177–215.

'Remarks on Philosophical Anthropology', in Horkheimer, *Between Philosophy and Social Science: Selected Early Writings* (Cambridge, MA: The MIT Press, 1993), pp. 151–75.

'The Rationalism Debate in Contemporary Philosophy', in Horkheimer, *Between Philosophy and Social Science: Selected Early Writings* (Cambridge, MA: The MIT Press, 1993), pp. 217–64.

'Reason Against itself: Some Remarks on Enlightenment', in *What is Enlightenment? Eighteenth-century Answers and Twentieth-century Questions*, ed. by James Schmidt (Berkeley–Los Angeles–London: University of California Press, 1996), pp. 359–67.

'Authority and the Family', in Horkheimer, *Critical Theory: Selected Essays* (New York: Continuum, 2002).

Dialectic of Enlightenment: Philosophical Fragments, trans. by Edmund F. N. Jephcott (Stanford, CA: Stanford University Press, 2002).

'Materialism and Metaphysics', in Max Horkheimer, *Critical Theory: Selected Essays*, trans. by Matthew J. O'Connell and others (New York: Continuum, 2002), pp. 10–46.

'Postscript', in *Critical Theory: Critical Theory: Selected Essays* (New York: Continuum, 2002), pp. 244–52.

'Traditional and Critical Theory', in Horkheimer, *Critical Theory: Selected Essays* (New York: Continuum, 2002), pp. 188–243.

A Life in Letters: Selected Correspondence, ed. and trans. Manfred R. Jacobson and Evelyn M. Jacobson (Lincoln, NE: University of Nebraska Press, 2007).

Theodor W. Adorno's collected works, letters and lecture courses

Gesammelte Schriften in 20 Bänden, ed. by Rolf Tiedemann, Gretel Adorno, Susan Buck-Morss and Klaus Schultz (Frankfurt am Main: Suhrkamp, 1970–1997).

Vol. 1: *Philosophische Frühschriften*, 2nd edition (Frankfurt am Main: Suhrkamp, 1990).
Vol. 2: *Kierkegaard. Konstruktion des Ästhetischen* (Frankfurt am Main: Suhrkamp, 2003).
Vol. 4: *Minima moralia. Reflexionen aus dem beschädigten Leben* (Frankfurt am Main: Suhrkamp, 2003).
Vol. 5: *Zur Metakritik der Erkenntnistheorie. Drei Studien zu Hegel*, 5th edition (Frankfurt am Main: Suhrkamp, 2021).
Vol. 6: *Negative Dialektik. Jargon der Eigentlichkeit*, 8th edition (Frankfurt am Main: Suhrkamp, 2018).
Vol. 7: *Ästhetische Theorie* (Frankfurt am Main: Suhrkamp, 1973).
Vol. 8: *Soziologische Schriften* I, 4th edition (Frankfurt am Main: Suhrkamp, 2018).
Vol. 9.1: *Soziologische Schriften* II.1, 3rd edition (Frankfurt am Main: Suhrkamp, 2019).
Vol. 9.2: *Soziologische Schriften* II.2, 3rd edition (Frankfurt am Main: Suhrkamp, 2019).
Vol. 10.1: *Kulturkritik und Gesellschaft I, Prismen – Ohne Leitbild*, 7th edition (Frankfurt am Main, Suhrkamp, 2018).
Vol. 10.2: *Kulturkritik und Gesellschaft II*: *Eingriffe – Stichworte – Anhang*, 7th edition (Frankfurt am Main: Suhrkamp, 2018).
Vol. 11: *Noten zur Literatur* (Frankfurt am Main: Suhrkamp, 2022).
Vol. 12: *Philosophie der neuen Musik* (Frankfurt am Main: Suhrkamp, 2003).
Vol. 14: *Dissonanzen. Einleitung in die Musiksoziologie*, 6th edition (Frankfurt am Main: Suhrkamp, 2003).
Vol. 17: *Musikalische Schriften* IV, *Moments musicaux. Impromptus* (Frankfurt am Main: Suhrkamp, 2003).
Vol. 20.1 *Vermischte Schriften I. Theorien und Theoretiker – Gesellschaft, Unterricht, Politik*, 5th edition (Frankfurt am Main: Suhrkamp, 2023).
Vol. 20.2 *Vermischte Schriften II. Aesthetica – Miscellanea*, 5th edition (Frankfurt am Main: Suhrkamp, 2023).

Erziehung zur Mündigkeit: Vorträge und Gespräche mit Hellmut Becker 1959-1969, ed. by Gerd Kadelbach (Frankfurt am Main: Suhrkamp, 1971).
Einleitung in die Soziologie, ed. by Christoph Gödde (Frankfurt am Main: Suhrkamp, 1993).
Frankfurter Adorno Blätter III, ed. by Theodor W. Adorno Archiv (Munich: editiontext+kritik, 1994).

'Individuum und Gesellschaft. Entwürfe und Skizzen', in *Frankfurter Adorno Blätter VIII*, ed. by Rolf Tiedemann (Munich: text + kritik, 2003), pp. 60–94.

'Contra Paulum', in Adorno, *Briefe und Briefwechsel*, Band II: 1938-1944, ed. by Christoph Gödde und Henri Lonitz (Frankfurt am Main: Suhrkamp, 2004), pp. 475–502.

Zur Lehre von der Geschichte und von der Freiheit (1964/1965), ed. by Rolf Tiedemann (Frankfurt am Main: Suhrkamp, 2006).

Probleme der Moralphilosophie, ed. by Thomas Schröder (Frankfurt am Main: Suhrkamp, 2010).

'Kultur und Culture' [7.6.1957], *Vorträge 1949-1968*, ed. by Michael Schwarz, *Nachgelassene Schriften*, Abteilung V: *Vorträge und Gespräche*, Vol. 1 (Berlin: Suhrkamp Verlag, 2019), pp. 156–76.

Theodor W. Adorno's works in English

'The actuality of philosophy', *Telos*, 31 (1977), 120–33.

'The Idea of Natural History', *Telos: Critical Theory of the Contemporary*, 60 (1984), 111–24.

Kierkegaard: Construction of the Aesthetic, trans. by Robert Hullot-Kentor (Minneapolis: University of Minnesota Press, 1989).

Hegel: Three Studies, trans. by Shierry Weber Nicholsen (Cambridge, MA: MIT Press, 1993).

Theodor W. Adorno & Hellmut Becker, 'Education for maturity and responsibility', *History of the Human Sciences*, 12.3 (1999), 21–34.

Introduction to Sociology, ed. by Christoph Godde, trans. by Edmund Jephcott (Stanford, CA: Stanford University Press, 2000).

Problems of Moral Philosophy, trans. by Rodney Livingstone (Cambridge: Polity Press, 2000).

Aesthetic Theory, ed. by Gretel Adorno and Rolf Tiedemann, newly translated, edited, and with a translator's introduction by Robert Hullot-Kentor (London and New York: Continuum, 2002).

Negative Dialectics, trans. by E. B. Ashton (New York: Taylor & Francis e-Library, 2004).

Critical Models: Interventions and Catchwords, trans. by Henry W. Pickford (New York: Columbia University Press, 2005).

Minima moralia; Reflections from Damaged Life, trans. by Edmund F. N. Jephcott (London: Verso, 2005).

History and Freedom: Lectures 1964–1965, edited by Rolf Tiedemann, trans. by Rodney Livingstone (Cambridge: Polity Press, 2006).

Philosophy of New Music, trans, edited, and with an introduction by Robert Hullot-Kentor (Minneapolis and London: University of Minnesota Press, 2006).

'Kultur and Culture', *Social Text*, 27.2/99 (2009), 145–158.

'Dossier: Adorno's Notes on Beckett', trans. by Dirk Van Hulle and Shane Weller, *Journal of Beckett Studies*, 19.2 (2010), 157–78.

Notes to literature, ed. by Rolf Tiedemann, trans. by Shierry Weber Nicholson (New York: Columbia University Press, 2019).

Other primary sources

Agamben, Giorgio, *The Open: Man and Animal*, trans. by Kevin Attell (Stanford, CA: Stanford University Press, 2004).

Beckett, Samuel, *Fin de partie* (Paris: Les Éditions de Minuit, 1957).

Beckett, Samuel, *Molloy, Malone Dies, The Unnamable* (London: Calder, 1994).

Beckett, Samuel, *The Complete Dramatic Works of Samuel Beckett* (London: Faber & Faber, 2006).

Benjamin, Walter, *The Origin of German Tragic Drama*, trans. by John Osborne (London: Verso, 1977).

Burckhardt, Jacob, *Die Kultur der Renaissance in Italien. Ein Versuch*, in Burckhardt, *Das Geschichtswerk in 2 Bänden*, I (Frankfurt am Main: Zweitausendeins, 2007).

Descola, Philippe, *Par-delà nature et culture* (Paris: Gallimard, 2005).

Descola, Philippe, *Beyond Nature and Culture*, trans. by Janet Lloyd (Chicago: The University of Chicago Press, 2013).

Fanon, Frantz, *Les damnés de la terre* (Paris: La Découverte, 2009).

Foucault, Michel, *Les mots et les choses* (Paris: Gallimard, 1990).

Foucault, Michel, *The Orders of Things: An Archeology of the Human Sciences* (New York: Vintage Books, 1994).

Foucault, Michel, 'L'homme est-il mort?', entretien avec Claude Bonnefoy (1966), in Michel Foucault, *Dits et écrits I, 1954-1975* (Paris: Gallimard, 2001), pp. 568–72.

Freud, Sigmund, *Civilization and Its Discontents*, in The Standard Edition of the Complete Psychological Works of Sigmund Freud, translated from the German under the general editorship of James Strachey, in collaboration with Anna Freud; assisted by Alix Strachey and Alan Tyson, Volume xxi (1927–1931): *The Future of an Illusion, Civilization and its Discontents and Other Works* (London: The Hogarth Press and Institute of Psycho-Analysis, 1961), pp. 64–145.

Fromm, Erich, 'Humanismus und Psychoanalyse', in Fromm, *Gesamtausgabe IX: Sozialistischer Humanismus und humanistische Ethik* (Munich: DTV, 1999), pp. 3–11.

Fromm, Erich, *The Sane Society* (London and New York: Routledge, 2002).

Fromm, Erich, *Escape from Freedom*, with a new foreword by S. Sloan (New York and Tokyo: Ishi Press, 2011).

Grassi, Ernesto, *Heidegger and the Question of Renaissance Humanism: Four Studies* (Binghamton: Center for Medieval and Early Renaissance Studies, 1983).

Hagen, Karl, *Deutschlands literarische und religiöse Verhältnisse im Reformationszeitalter. Mit besonderer Rücksicht auf Wilibald Pirkheimer*, 3 vols (Erlangen: Palm, 1841–1844).

Hegel, Georg Wilhelm Friedrich, *Philosophy of Mind*, trans by William Wallace and Arnold V. Miller, revisions and commentary by Michael J. Inwood (Oxford: Clarendon Press, 2007).

Husserl, Edmund, *The Crisis of European Sciences and Transcendental Phenomenology*, trans. by David Carr (Evanston, Illinois: Northwestern University Press, 1970).

Korsch, Karl, *Marxism and Philosophy*, trans. by Fred Halliday (London: Verso, 2012).

Latour, Bruno, *Face à Gaïa. Huit conférences sur le nouveau régime climatique* (Paris: La Découverte, 2015, 2023).

Latour, Bruno, *Facing Gaia: Eight Lectures on the New Climatic Regime* (Cambridge: Polity Press, 2017).

Latour, Bruno, *Down to Earth: Politics in the New Climatic Regime* (Cambridge: Polity, 2018).

Lefebvre, Henri, *Dialectical Materialism*, trans. by John Sturrock (London: Jonathan Cape, 1968).

Lévi-Strauss, Claude, 'Jean-Jacques Rousseau, fondateur des sciences de l'homme', in Claude Lévi-Strauss, *Anthropologie structurale deux* (Paris: Plon, 1996 [1962, 1973]), pp. 45–56.

Lévi-Strauss, Claude, 'Jean-Jacques Rousseau, Founder of the Sciences of Man', in Claude Lévi-Strauss, *Structural Anthropology*, Volume 2, trans. by Monique Layton (Chicago: University of Chicago Press, 1983), pp. 33–43.

Lévi-Strauss, Claude, 'Race et culture', in Lévi-Strauss, *Le regard éloigné* (Paris: Plon, 2001 [1983]), pp. 24–48.

Lovelock, James, *The Revenge of Gaia: Earth's Climate in Crisis and the Fate of Humanity*, foreword by Sir Crispin Tickell (London: Penguin Books, 2007).

Lukács, Georg, 'The Ideology of Modernism', in Georg Lukács, *The Meaning of Contemporary Realism*, ed. and trans. by John Mander and Necke Mander (London: Merlin, 1963), pp. 17–46.

Lukács, Georg, *Die Theorie des Romans: Ein geschichtsphilosophischer Versuch über die Formen der großen Epik*, 9th Edition (Darmstadt: Luchterhand, 1971).

Lukács, Georg, *The Theory of the Novel: A Historico-Philosophical Essay on the Forms of Great Epic Literature*, trans. by Anna Bostock (Cambridge, MA: The MIT Press 1971).

Lukács, Georg, *History and Class Consciousness: Studies in Marxist Dialectics*, trans. by Rodney Livingstone (Cambridge, MA: The MIT Press, 1971).

Lukács, Georg, *Geschichte und Klassenbewußtsein. Studien über marxistische Dialektik* (Bielefeld: Aisthesis Verlag, 2013).

Marcuse, Herbert, 'Über den affirmativen Charakter der Kultur', in Marcuse, *Kultur und Gesellschaft*, Vol. 1 (Frankfurt am Main: Suhrkamp, 1965), pp. 56–102.

Marcuse, Herbert, 'The Affirmative Character of Culture', in Marcuse, *Negations: Essays in Critical Theory*, with translations from the German by J. J. Shapiro (London: MayFlyBooks, 2009), pp. 65–98.

Marck Siegfried, *Der Neuhumanismus als politische Philosophie* (Zürich: Aufbruch, 1938).

Marx, Karl, *Economic and Philosophic Manuscripts of 1844*, trans. by Martin Milligan (New York: Prometheus Book, 1988).

Marx, Karl and Friedrich Engels, *Die Heilige Familie oder Kritik der kritischen Kritik Gegen Bruno Bauer und Konsorten*, in Marx and Engels, *Werke*, hrsg. vom Institut für Marxismus-Leninismus, beim Zentral-Komitee der sozialistischen Einheitspartei Deutschlands, Vol. 2 (Berlin: Dietz Verlag, 1962), pp. 3–223.

Marx, Karl, *The Holy Family, or Critique of Critical Criticism. Against Bruno Bauer and Company*, in Marx and Engels, *Collected Works*, 12 vols, Vol. 4, 1844–45 (London: Lawrence & Wishart, 2010), pp. 5–211.

Marx, Karl, and Friedrich Engels, *Die deutsche Ideologie*, in Marx and Engels, *Werke*, vol. 3 (Berlin: Dietz Verlag, 1978).

Marx, Karl, and Friedrich Engels, *The German Ideology*, in Marx and Engels, *Collected Works*, Vol. 5, trans. by W. Lough, C. Dutt and C. P. Magill (London: Lawrence & Wishart, 1974, Electric Book, 2010).

Montaigne, Michel Eyquem de, *Les Essais*, ed. by Pierre Villey and Verdun-Louis Saulnier (Paris: Presses Universitaires de France, 2004).

Montaigne, Michel Eyquem de, *The Complete Essays of Michel de Montaigne*, trans. by Donald M. Frame (Stanford: Stanford University Press, 1948, 1958).

Niethammer, Friedrich Immanuel, *Der Streit des Philanthropinismus und Humanismus in der Theorie des Erziehungs-Unterrichts unsrer Zeit* (Jena: Frommann, 1808).

Nietzsche, Friedrich, *Human, All Too Human*, trans. by R. J. Hollingdale, with an introduction by Richard Schacht (Cambridge: Cambridge University Press, 1996).

Nietzsche, Friedrich, *On the Genealogy of Morality*, ed. by Keith Ansell-Pearson, trans. by Carol Diethe (Cambridge and New York: Cambridge University Press, 2006).

Quesnay, François, 'Des sciences et des arts', *Éphémérides du citoyen, ou Bibliothèque raisonnée des sciences morales et politiques*, 1.16 (1765), 241–56.

Sartre, Jean-Paul, 'Réponse à Albert Camus', in *Les Temps modernes*, 82 (August 1952), in Sartre, *Situations, IV* (Paris: Gallimard, 1964), pp. 90–125.

Sartre, Jean-Paul, 'Préface', in Frantz Fanon, *Les damnés de la terre* (Paris: La Découverte, 2009), pp. 17–36.

Schopenhauer, Arthur, *Prize Essay On the Basis of Morals*, trans. by Christopher Janaway, in Schopenhauer, *The Two Fundamental Problems of Ethics* (1841/1860) (Cambridge: Cambridge University Press, 2009), pp. 113–258.

Schopenhauer, Arthur, *The World as Will and Representation*, vol. 1 (1818/1844/1859), trans. and ed. by Christopher Janaway, Judith Norman, and Alistair Welchman (Cambridge: Cambridge University Press, 2014).

Tillich, Paul, 'Man and Society in Religious Socialism', in Tillich, *Main Works / Hauptwerke*, Volume 3 / Band 3, *Writings in the Social Philosophy and Ethics / Sozialphilosophische und ethische Schriften*, ed. by Carl H. Ratschow (Berlin: De Gruyter, 1998), pp. 487–502.

Voigt, Georg, *Die Wiederbelebung des classischen Alterthums oder das erste Jahrhundert des Humanismus*, 2 vols. (Berlin: Reimer 1859).

Weber, Max, *Die protestantische Ethik und der Geist des Kapitalismus*, in Weber, *Gesammelte Aufsätze zur Religionssoziologie* (Tübingen: J. C. B. Mohr, 1988), vol. 1, pp. 17–206.

Weber, Max, *The Protestant Ethic and the Spirit of Capitalism*, trans. by Talcott Parsons, introduction by Anthony Giddens (London and New York: Routledge, 2005).

Weber, Max, 'Wissenschaft als Beruf', in Weber, *Gesammelte Aufsätze zur Wissenschaftslehre* (Tübingen: J. C. B Mohr (Paul Siebeck), 1968).

Secondary Literature

Abromeit, John, *Max Horkheimer and the Foundations of the Frankfurt School* (New York: Cambridge University Press, 2011).

Altman Matthew C. and Cynthia D. Coe, 'Wolves, Dogs, and Moral Geniuses: Anthropocentrism in Schopenhauer and Freud', in *The Palgrave Schopenhauer Handbook*, ed. by Sandra Shapsay (Cham, Switzerland: Palgrave Macmillan, 2017), pp. 447–71.

Antonio, Robert J., 'Immanent Critique as the Core of Critical Theory: Its Origins and Developments in Hegel, Marx and Contemporary Thought', *The British Journal of Sociology*, 32.3 (1981), 330–45.

Bagla, Lusin, 'La nature entre changement climatique et séquestration du CO_2. De la recherche au pragmatisme', *L'Art du comprendre*, 21 (2012), 203–17.

Bégot, Jacques-Olivier, 'Critique et interprétation dans les premiers écrits philosophiques d'Adorno', in Theodor W. Adorno, *L'actualité de la philosophie et autres essais*, ed. by Jacques-Olivier Bégot (Paris: Éditions Rue d'Ulm, 2018), pp. 71–103.

Besnier, Jean-Michel, 'Un humanisme paradoxal', *Le Télémaque*, 21 (2002), 97–106.

Bödeker, Hans Erich, 'Menschheit, Humanismus, Humanität', in *Geschichtliche Grundbegriffe: Historisches Lexikon zur politisch-sozialen Sprache in Deutschland*, ed. by Otto Brunner, Werner Conze, and Reinhart Koselleck (Stuttgart: Klett-Cotta, 1982), vol. 3, pp. 1063–1128.

Boer, Roland, 'A Totality of Ruins: Adorno on Kierkegaard', *Cultural Critique*, 83 (2013), 1-30.

Bollenbeck, Georg, *Bildung und Kultur: Glanz und Elend eines deutschen Deutungsmusters* (Frankfurt am Main: Insel, 1994).

Bommel, Sebastiaan van, *Classical Humanism and the Challenge of Modernity: Debates on Classical Education in 19th-Century Germany* (Berlin: De Gruyter, 2015).

Borman, David A., 'Materialism in Critical Theory: Marx and the Early Horkheimer', in *The Palgrave Handbook of Critical Theory*, ed. by Michael J. Thompson (New York: Palgrave Macmillan, 2017), pp. 207–29.

Bosakova Kristina, and Martina F. Bykova, 'Hegel and Niethammer on the Educational Practice in Civil Society', *Journal of Philosophy of Education*, 55.1 (2021), 99–125.
Brague, Rémi, *Le propre de l'homme* (Paris: Flammarion, 2015).
Bryson, Michael, 'Introduction. The Old Argument: Humanism and Anti-Humanism', in *The Routledge Companion to Humanism and Literature*, ed. by Michael Bryson (New York: Routledge, 2022) pp. 1–9.
Buck-Morss, Susan, *The Origin of Negative Dialectics* (New York: The Free Press, 1977).
Butler, Judith, *Precarious Life: The Powers of Mourning and Violence* (London: Verso, 2004).
Cacciatore, Giuseppe, 'Nuovo umanesimo e filosofia interculturale', *Humanitas*, 69.4–5 (2014), 584–95.
Carbone, Raffaele, 'Umanesimo, scetticismo, società borghese: Horkheimer lettore di Montaigne', *Atti dell'Accademia delle Scienze morali e politiche*, 128 (2018), 7–40.
Carbone, Raffaele, 'Società borghese, umanesimo e teoria critica nella prospettiva di Max Horkheimer', *Archivio di Storia della Cultura*, 33 (2020), 245–72.
Carbone, Raffaele, 'Formes de vie et dynamique historique chez Max Horkheimer', *Archives de Philosophie*, 85.2 (2022), 45–60.
Cerutti, Furio, 'Marxismo e sociologia nella Repubblica federale tedesca', *Belfagor*, 24.6 (1969), pp. 633–52.
Cook, Deborah, *Adorno on Nature* (London and New York: Routledge, 2014).
Cook, Deborah, *Adorno, Foucault and the Critique of the West* (London and New York, 2018).
Cortella, Lucio, *Per una dialettica nella finitezza. Adorno e il programma di una dialettica negativa* (Rome: Meltemi, 2006).
Dahbour, Omar, 'Totality, Reason, Dialectics: The Importance of Hegel for Critical Theory from Lukács to Honneth', in *The Palgrave Handbook of Critical Theory*, ed. by Michael J. Thompson (New York: Palgrave Macmillan, 2017), pp. 87–108.
Demirović, Alex, 'Raison et émancipation', in *La dialectique de la raison. Sous bénéfice d'inventaire*, ed. by Katia Genel (Paris: Éditions de la Maison des Sciences de l'Homme, 2017), pp. 87–104.
Denham, Helen, 'The Cunning of Unreason and Nature's Revolt: Max Horkheimer and William Leiss on the Domination of Nature', *Environment and History*, 3.2, *Ecological Visionaries/Ecologised Visions* (June, 1997), 149–75.

Desrosiers, Martin, 'Un humanisme négatif. Horkheimer, Adorno et la critique du "kitsch humaniste"', in *Les Normes et le possible. Héritage et perspectives de l'École de Francfort*, ed. by Pierre-François Noppen, Gérard Raulet and Iain Macdonald (Paris: Éditions de la Maison des sciences de l'homme, 2012), pp. 285–307.

Distaso, Leonardo, *Marcuse, Adorno. Percorsi tra estetica e politica* (Rome: Carocci, 2022).

Epstein, Mikhail, *The Transformative Humanities: A Manifesto* (New York: Bloomsbury, 2012).

Falchi Pellegrini, Maria Antonietta, 'Il Montaigne di Horkheimer', *Il pensiero politico*, 22.2 (1989), 248–63.

Feenberg, Andrew, 'Why Students of the Frankfurt School Will Have to Read Lukács', in *The Palgrave Handbook of Critical Theory*, ed. by Michael J. Thompson (New York: Palgrave Macmillan, 2017), pp. 109–33.

Ferrarese, Estelle, *The Fragility of Concern for Others: Adorno and the Ethics of Care*, trans. by Steven Corcoran (Edinburgh: Edinburgh University Press, 2020).

Finlayson, James Gordon, 'Adorno's Metaphysics of Moral Solidarity in the Moment of its Fall', in *A Companion to Adorno*, ed. by Peter E. Gordon, Espen Hammer and Max Pensky (Hoboken, NJ: Wiley Blackwell, 2020), pp. 615–30.

Fischbach, Franck, *Pour la Théorie critique. Raison, nature et société* (Paris: Vrin, 2024).

Flipo, Fabrice, *Les écologistes sans la nature. Le faux ami Bruno Latour* (Paris: Textuel, 2025).

Flodin, Camilla, 'In the Name of the Rhinoceros: Expression beyond Human Intention', in *Adorno's Rhinoceros: Art, Nature, Critique*, edited by Antonia Hofstätter and Daniel Steuer (London: Bloomsbury, 2022), pp. 19–42.

Genel, Katia, *Autorité et émancipation. Horkheimer et la Théorie critique* (Paris: Payot, 2013).

Genel, Katia, 'Une maladie de la raison? La métaphore médicale dans la théorie critique de Horkheimer', in *Genealogie della modernità: prospettive critiche*, ed. by Raffaele Carbone and Luca Scafoglio (Naples: FedOA Press, 2024).

Grivaux, Agnès, 'Usages de la non-identité dans la dialectique nature-histoire chez Adorno et Horkheimer. À propos de la différence entre besoin et pulsion', *Trajectoires*, Hors série, 4 (2020), Online erschienen am: 20 Januar 2020, abgerufen am 26 Februar 2026. URL: http://journals.

openedition.org/trajectoires/4142; DOI: https://doi.org/10.4000/trajectoires.4142.

Grivaux, Agnès, 'Natur zwischen Übermacht, Unschuld und Instrument, Zur Skizze eines minimalen Naturbegriffs bei Adorno', *Deutsche Zeitschrift für Philosophie*, 69.6 (2021), 989–1002.

Guérard de Latour, Sophie, 'L'humanisme, une valeur à partager entre différentes cultures. À quelles conditions', *L'Observatoire*, 49.1 (2017), 25–8.

Habermas, Jürgen, 'The Entwinement of Myth and Enlightenment: Re-Reading Dialectic of Enlightenment', trans. by Thomas Y. Levin, *New German Critique*, 26 (Spring–Summer, 1982), *Critical Theory and Modernity*, 13–30.

Hakkarainen, Heidi, 'Contagious Humanism in Early Nineteenth-Century German-Language Press', *Contributions to the History of Concepts*, 15.1 (2020), 22–46.

Hartog, François, 'Lettre sur l'Humanisme et les Humanités', *Anabases. Traditions et réceptions de l'Antiquité*, Études sur la circulation et la réception des savoirs offertes à Pascal Payen. Actualités et débats, 35 (2022), 227–43.

Hartog, François, *Départager l'humanité. Humains, humanismes, inhumains* (Paris: Gallimard, 2024).

Hesse, Heidrun, *Vernunft und Selbstbehauptung. Kritische Theorie als Kritik der neuzeitlichen Rationalität* (Frankfurt am Main: Fischer Taschenbuch, 1984).

Hinchman, Lewis P., 'Is Environmentalism a Humanism?', *Environmental Values*, 13.1 (2004), 3–29.

Hoffmann, Arnd, 'Adorno und die Tiere', in *Adorno-Handbuch: Leben – Werk – Wirkung*, ed. by Richard Klein, Johann Kreuzer and Stefan Müller-Doohm, 2nd edition (Berlin: J. B. Metzler, 2019), pp. 444–53.

Holt, Matthew, 'Catastrophe, autonomy and the future of modernism: Trying to understand Adorno's reading of *Endgame*', *Samuel Beckett Today – Aujourd'hui*, vol. 14 (2004), 260–75.

Honneth, Axel, *The Fragmented World of the Social: Essays in Social and Political Philosophy* (Albany: State University of New York Press, 2005).

Hullot-Kentor, Robert, 'Introduction to T. W. Adorno's "The Idea of Natural-History"', in Hullot-Kentor, *Things Beyond Resemblance: Collected Essays on Theodor W. Adorno* (New York: Columbia University Press, 2006), pp. 234–51.

Jaeggi, Rahel, '"No Individual Can Resist": *Minima Moralia* as Critique of Forms of Life', trans. by James Ingram, *Constellations*, 12.1 (2005), 65–82.

Jakelić, Slavica, 'Humanism and Its Critics', in *The Oxford Handbook of Humanism*, ed. by Anthony B. Pinn (Oxford: Oxford University Press, 2021), pp. 265–93

Jameson, Fredric, *Late Marxism: Adorno, or, the Persistence of the Dialectic* (London: Verso, 1990).

Jarvis, Simon, *Adorno: A Critical Introduction* (New York: Routledge, 1998).

Jay, Martin, *The Dialectical Imagination: A History of the Frankfurt School and the Institute of Social Research, 1923-1950* (Boston, MA: Little, Brown and Co., 1973).

Jay, Martin, *Marxism and Totality: The Adventures of a Concept from Lukács to Habermas* (Berkeley, CA: University of California Press, 1984).

Kämper, Felix, 'Domination by nature: On an unresolved tension in the Frankfurt School', *Thesis Eleven*, 2025, 1–16.

Kozlarek, Oliver, 'From the humanism of critical theory to critical humanism', *European Journal of Social Theory*, 24.2 (2020), 1–18.

Kozlarek, Oliver, *The Critical Humanism of the Frankfurt School as Social Critique* (Lanham, Maryland: Lexington Books, 2024).

Larrère, Catherine, and Raphaël Larrère, *Du bon usage de la nature. Pour une philosophie de l'environnement* (Paris: Flammarion, 2022; Paris: Aubier, 1997).

Lemm, Vanessa, 'Critical theory and affirmative biopolitics: Nietzsche and the domination of nature in Adorno/Horkheimer', *Journal of Power*, 3.1 (2010), 75–95.

Liotta, Daniel, 'Foucault: Une introduction à une critique de l'humanisme', *Humanisme*, 293.3 (2011), 35–45.

Lohmann, Georg, 'The Failure of Self-Realization: An Interpretation of Horkheimer's *Eclipse of Reason*', in *On Max Horkheimer: New Perspectives*, ed. by Seyla Benhabib, Wolfgang Bonß, and John McCole (Cambridge, MA–London: The MIT Press, 1993), pp. 387–412.

McKanan, Dan, 'Humanism and Environmentalism', in *The Oxford Handbook of Humanism*, ed. by Anthony B. Pinn (New York, NY: Oxford University Press, 2021), pp. 509–30.

Merchant, Carolyn, 'Francis Bacon and the "vexations of art": Experimentation as intervention', *The British Journal for the History of Science*, 46.4 (2013), 551–99.

Moir, Cat, 'Second Nature and the Critique of Ideology in Hegel and the Frankfurt School', in *Hegel and the Frankfurt School*, ed. by Paul Giladi (New York – Abingdon, Oxon: Routledge, 2021), pp. 115–39.

Morgan, Alastair, 'Reconciliation with Nature: Adorno on Reason, Nature and Critique', *Adorno Studies*, 3.1 (2019), 20–32.
Moutot, Gilles, *Essai sur Adorno* (Paris: Payot, 2010).
Moutot, Gilles, '"Le souvenir de la nature dans le sujet". Une actualité de La Dialectique de la raison', *Astérion*, 7 (2010), https://journals.openedition.org/asterion/1622.
Murphy, Ann V., 'Corporeal Vulnerability and the New Humanism', *Hypatia*, 26.3 (2011), 575–90.
Nelson, Eric S., 'Revisiting the Dialectic of Environment: Nature as Ideology and Ethics in Adorno and the Frankfurt School', *Telos*, 155 (Summer 2011): 105-126.
Nilsson, Alice, 'The humanism of critical theory: The Frankfurt School's "realer humanismus"', *Philosophy & Social Criticism*, 51.5 (2023), 727–46.
Noppen, Pierre-François, 'The Anthropology in *Dialectic of Enlightenment*' in *A Companion to Adorno*, ed. by Peter E. Gordon, Espen Hammer and Max Pensky (Hoboken, NJ: Wiley Blackwell, 2020), pp. 207–20.
Norberg, Jakob, 'Adorno's Advice: *Minima Moralia* and the Critique of Liberalism', *Publications of the Modern Language Association* (*PMLA*), 126.2 (2011), 398–411.
Oberle, Eric, *Theodor Adorno and the Century of Negative Identity* (Stanford, CA: Stanford University Press, 2018).
Ombrosi, Orietta, 'Le *miserere* des bêtes. Max Horkheimer et Theodor W. Adorno face à l'animalité', *Lignes*, 28.1 (2009), 146–66.
Omosulu Rotimi, and Terzungwe Inja, 'Humanism and Environmentalism: A Dialogue', *Nnamdi Azikiwe Journal of Philosophy*, 11.2 (2019), 87–100.
Parra-Ayala, Andrés Felipe, 'Materialismo ecológico como materialismo de la reproducción: un diálogo entre Theodor Adorno y Raquel Gutiérrez Aguilar', *Revista de Estudios Sociales*, 91 (2025), 23–38.
Pensky, Max, 'Natural History: the Life and Afterlife of a Concept in Adorno', *Critical Horizons*, 5 (2004), 227–58.
Peyrical, Aurélia, 'T. W. Adorno, H. Becker, and the Challenges of Education in an "Administered World" (1955–1969): Unpublished Radio Conversations from the Theodor W. Adorno Archive', in *The Archives of Critical Theory*, ed. by Isabelle Aubert and Marcos Nobre (Cham, Switzerland: Springer, 2023), pp. 165–89.
Petrucciani, Stefano, *Ragione e dominio: L'autocritica della razionalità occidentale in Adorno e Horkheimer* (Rome: Salerno editrice, 1984).

Petrucciani, Stefano, *Theodor W. Adorno's Philosophy, Society, and Aesthetics* (Cham: Palgrave Macmillan, 2021).
Petrucciani, Stefano, *La Scuola di Francoforte* (Rome: Carocci, 2023).
Pilbeam, Pamela, 'Bourgeois society', in *A Companion to Nineteenth-Century Europe: 1789-1914*, ed. by Stefan Berger (Oxford–Malden, MA: Blackwell Publishing Ltd, 2006), pp. 86–97.
Plet, Laurent, *Essai sur la* Dialectique négative *d'Adorno: matérialisme critique et utopie* (Paris, Classiques Garnier, 2016).
Plummer, Ken, *Critical Humanism: A Manifesto for the 21st Century* (Cambridge: Polity Press, 2021).
Ponsetto, Antonio, *Max Horkheimer. Dal mito della distruzione alla distruzione del mito* (Bologna: il Mulino, 1981).
Post, Werner, *Kritische Theorie und metaphysischer Pessimismus. Zum Spätwerk Max Horkheimers* (München: Kösel Verlag, 1971).
Postone, Moishe, *Time, Labor, and Social Domination: A Reinterpretation of Marx's Critical Theory* (Cambridge: Cambridge University Press, 2003).
Raga, Vicente, 'Tentativas sobre Montaigne: Horkheimer y la función del escepticismo', *Estudios de Filosofía*, 55 (2017), 82–102.
Reeves, Craig, 'Causality and Critical Theory: Nature's Order in Adorno, Cartwright and Bhaskar', *Journal of Critical Realism*, 8.3 (2009), 316–342.
Reitemeyer, Ursula, *Umbruch in Permanenz. Eine Theorie der Moderne zwischen Junghegelianismus und Frankfurter Schule* (Münster: Waxmann, 2007).
Renault, Emmanuel, 'Sujet-objet: le dispositif Hegel-Kant', *Cahiers philosophiques*, 154.3 (2018), 9–28.
Rocco, Christopher, 'Between Modernity and Postmodernity: Reading *Dialectic of Enlightenment* against the Grain', *Political Theory*, 22.1 (1994), 71–97.
Rose, Gillian, *The Melancholy Science: An Introduction to the Thought of Theodor W. Adorno* (London: Verso, 2014; London: Macmillan Press, 1978).
Roulx, Dominic, 'Le concept de nature chez Adorno', *Laval théologique et philosophique*, 79.3 (2023), 435–49.
Rüsen, Jörn, *Humanism: Foundations, Diversities, Developments* (London and New York: Routledge, 2021).
Said, Edward W., *Humanism and Democratic Criticism* (New York: Columbia University Press, 2004).
Sanbonmatsu, John, 'Introduction', in *Critical Theory and Animal Liberation*, ed. by John Sanbonmatsu (Plymouth: Rowman & Littlefield, 2011), pp. 1–32.

Scafoglio, Luca, *Il disagio dell'astrazione. Una lettura a partire da Adorno* (Naples: La Città del Sole, 2021).

Schmid Noerr, Gunzelin, *Das Eingedenken der Natur im Subjekt. Zur Dialektik von Vernunft und Natur in der Kritischen Theorie Horkheimers, Adornos und Marcuses* (Darmstadt: Wissenschaftliche Buchgesellschaft, 1990).

Schmidt, Alfred, *Die Kritische Theorie als Geschichtsphilosophie* (München: Hanser, 1976).

Schmidt, Alfred, *Kritische Theorie, Humanismus, Aufklärung. Philosophische Arbeiten 1969-1979* (Stuttgart: Philipp Reclam Jun., 1981).

Schmidt, Alfred, 'Aufklärung und Mythos im Werk Max Horkheimers', in *Max Horkheimer heute: Werk und Wirkung*, ed. by Alfred Schmidt and Norbert Altwicker (Frankfurt am Main: Fischer Taschenbuch Verlag, 1986), pp. 180–243.

Schmidt, Alfred, *The Concept of Nature in Marx*, trans. by Ben Fowkes (London–New York: Verso, 2014).

Schmidt, Alfred, and Gian Enrico Rusconi, *La Scuola di Francoforte: Origini e significato attuale* (Bari: De Donato, 1972).

Schultz, Markus, 'Siegfried Marck: Politische Biographie eines jüdisch-intellektuellen Sozialdemokraten' (PhD dissertation, University of Göttingen, 2016).

Schweppenhäuser, Gerhard, 'Nominalismuskritik und "negative Metaphysik". Philosophiegeschichtliche Überlegungen zum Begriff der Natur in der Dialektik der Aufklärung', in *Zur Kritik der regressiven Vernunft. Beiträge zur 'Dialektik der Aufklärung'* ed. by Gunzelin Schmid Noerr and Eva-Maria Ziege (Wiesbaden: Springer, 2019), pp. 65–89.

Schweppenhäuser, Gerhard, '"Jenseits der ästhetischen Immanenz". Natur und Gesellschaft in Adornos Theorie des Ästhetischen', *Scenari*, 14 (2021), 239–58.

Schweppenhäuser, Gerhard, *Zur Aktualität von Max Horkheimer. Einführung in sein Werk* (Wiesbaden: Springer, 2023).

Shapshay, Sandra, 'Schopenhauer on the Moral Considerability of Animals: Toward a Less Anthropocentric Ethics', in *The Palgrave Schopenhauer Handbook*, ed. by Sandra Shapsay (Cham, Switzerland: Palgrave Macmillan, 2017), pp. 283–98.

Stengel, Friedemann, *Genealogie des Humanismus. Debatten – Kritik – Neue Perspektiven* (Bielefeld: transcript Verlag, 2024).

Stone, Alison, 'Adorno and the Disenchantment of Nature', *Philosophy and Social Criticism*, 2006, 32.2: 231–53.

Taleb, Mohammed, 'Psyché, cosmos et âme du monde. Fragments d'histoire de l'écopsychologie', *L'Art du comprendre*, 21, *La nature et son souci. Philosophie et* écologie (2012), 187–201.

Taylor, Charles, *A Secular Age* (Cambridge, MA: The Belknap Press of Harvard University Press, 2007).

Todte, Mario, *Georg Voigt* (*1827–1891*): *Pionier der historischen Humanismusforschung* (Leipzig: Leipziger Universitätsverlag, 2004).

Toledo, Camille de, *Une histoire du vertige* (Lagrasse: Verdier, 2023).

Tomba, Massimiliano, 'Adorno's Account of the Anthropological Crisis and the New Type of Human', in *(Mis)readings of Marx in Continental Philosophy*, ed. by Jernej Habjan and Jessica Whyte (London: Palgrave Macmillan, 2014), pp. 34–50.

Toussaint, Stéphane, *Humanismes Antihumanismes: De Ficin à Heidegger*, I (Paris: Les Belles lettres, 2008).

Vogel, Steven, *Against Nature: The Concept of Nature in Critical Theory* (Albany: State University of New York Press, 1996).

Vuillerod, Jean-Baptiste, *Theodor W. Adorno. La domination de la nature* (Paris: Éditions Amsterdam, 2011).

Weber, Tomas, 'Metaphysics of the Common World: Whitehead, Latour, and the Modes of Existence', *The Journal of Speculative Philosophy*, 30.4 (2016), 515–33.

Weisberg, Zipporah, 'Animal Repression: Speciesism as Pathology', in *Critical Theory and Animal Liberation*, ed. by John Sanbonmatsu (Plymouth: Rowman & Littlefield, 2011), pp. 177–193.

Weller, Shane, 'Adorno's Notes on *The Unnamable*', *Journal of Beckett Studies*, 19.2, *Beckett and Germany* (2010), 179–195.

Wiggershaus, Rolf, *The Frankfurt School: Its History, Theories, and Political Significance*, trans. by Michael Robertson (Cambridge, MA: The MIT Press, 1995).

Wiggershaus, Rolf, *Max Horkheimer zur Einführung* (Hamburg: Junius Verlag, 1998).

Wilke, Sabine, 'Enlightenment, Dialectic, and the Anthropocene: Bruised Nature and the Residues of Freedom', *Telos*, 177 (2016), 83–106.

Wolff, Francis, *Plaidoyer pour l'universel. Fonder l'humanisme* (Paris: Arthème Fayard, 2019, 2021).

Zepp, Susanne, 'Max Horkheimer: Montaigne und die Funktion der Skepsis', in *Textgelehrte. Literaturwissenschaft und literarisches Wissen im Umkreis der Kritischen Theorie*, ed. by Nicolas Berg and Dieter Burdorf (Göttingen: Vandenhoeck & Ruprecht, 2014), pp. 81–90.

INDEX

MIMESIS GROUP
www.mimesis-group.com

MIMESIS INTERNATIONAL
www.mimesisinternational.com
info@mimesisinternational.com

MIMESIS EDIZIONI
www.mimesisedizioni.it
mimesis@mimesisedizioni.it

ÉDITIONS MIMÉSIS
www.editionsmimesis.fr
info@editionsmimesis.fr

MIMESIS COMMUNICATION
www.mim-c.net

MIMESIS EU
www.mim-eu.com